INTRODUCTION

The Fast Facts Encyclopedia of Science and Technology will amaze you with its range of fascinating topics. You will find out how rocket engines are powered, how your brain works, and everything you need to know about aliens, bows and arrows, and submarines!
This encyclopedia has a user-friendly contents and a clearly defined index to help you find out all you can – everyone you know will be surprised and delighted by your scientific and technological knowledge!

General Science
What is time and why do we measure it in the way we do? What are you seeing when you look in the mirror, and why does this happen? How does a light bulb work, and who was the first person to discover electricity? The answers to all of these questions are in this chapter.

Energy
Have you ever wondered how rockets are powered, why you need to eat, or why we all need the Sun in order to live? Find out in this chapter!

THE
FAST FACTS
ENCYCLOPEDIA

OF

SCIENCE
& TECHNOLOGY

© Aladdin Books Ltd 2000

Designed and produced by
Aladdin Books Ltd
28 Percy Street
London W1P 0LD

First published in the United States by
Copper Beech Books, an imprint of
The Millbrook Press
2 Old New Milford Road
Brookfield, Connecticut 06804

ISBN: 0-7613-0930-6

Printed in Italy

Some of the material in this book was
previously published in other Aladdin Books series.

THE
FAST FACTS
ENCYCLOPEDIA
OF

S C I E N C E
& TECHNOLOGY

Copper Beech Books
Brookfield, Connecticut

CONTENTS

Machines on Land and in the Air

Why not make your own hot-air balloon? Learn how to do this, and find the answers to any questions you have about planes and helicopters, such as what if there were no pilot? You can also find out about motorcycles, bicycles, and sporting trucks, and discover who were the very first people to invent the car!

Human Body

Do you know everything there is to know about your body? You'll see what the inside of your eye looks like, and find out the important job that blood does when it pumps around your body. You can also discover how we manage to breathe without even trying!

Arms and Armor

How will a pilot escape if the plane is attacked by the enemy? How fast can fighting ships travel? The topics covered by this chapter will answer these questions as well as providing exciting information about tanks, submarines, and more!

Inventions

Who invented the microscope? What is virtual reality? You can find out who the inventors were and how they succeeded when you read about inventions such as the camera, the telephone, and the microchip.

Stars, Planets, and Space

The Milky Way and UFOs may have seemed like mysteries until you read about them in this chapter – you will also find the answers to your strangest questions, such as what if the universe started to shrink?

This encyclopedia will take you on an exciting journey through science and technology presenting you with lots of facts in a fun and interesting way.

COLORED LIGHT

Color and light are inseparable. The eye perceives color because of three sets of receptors called cones, one for each of the primary colors. All the colors seen on a television screen are made up from colored dots of the three primary colors. The color of an object will depend on which colors of the spectrum it absorbs, and which it reflects. An object looks yellow because the yellow light rays are reflected into the eye, while the rest are absorbed. A black object absorbs all the colors of the spectrum, while a white object reflects them all.

BRIGHT IDEAS

Reconstitute white light by spinning a wheel like the one shown here. Color the wheel with colors of the spectrum and then spin it.

Make a hole in one end of a shoe box. Remove the lid and fill the box with objects of different colors. Make 2 or 3 cellophane covers, each a different color. Cover it with one of them, then shine a flashlight through the hole. Look through the cellophane. What color are the objects?

Colors of
the spectrum

The colors
disappear

SPOTLIGHTS

1 and 2. To mix colored light you will need cellophane filters in the three primary colors – green, red, and blue. Attach these filters to three long, cardboard tubes. Make sure the filters cover the end of each tube completely.

1

WHY IT WORKS

We have discovered by splitting light that each color has a different wavelength. Mixing colored lights produces new colors by adding light of different wavelengths. Colored light mixtures are sometimes called additive color mixtures or color by addition. Luminous sources of light, like color televisions, combine colors by mixing very small dots of light. Black means the absence of light because there are no colors to mix together. When red, green, and blue lights are combined, white light is the result. A secondary color is an equal mixture of two primary colors. Red and green lights shining onto a white object will make it appear to be yellow. Any two colors of light that form white light when mixed are called complementary. Other colors are formed by mixing the primary colors in different proportions.

3. Place a large sheet of white oak tag on the floor in a darkened room. Three people need to hold the tubes at right angles to the floor while shining a flashlight into the top of each. The beams of light should be directed onto the white oak tag.

4. Find out how to make new colors by positioning the three lights in such a way as to produce a variety of combinations. Where the three combine, white light is produced.

WHAT IS ENERGY?

Energy is the ability to do things. Without it we couldn't get up in the morning, or turn on the lights, or drive the car. Plants wouldn't grow, the rain wouldn't fall, the Sun wouldn't shine. Everything we do needs a supply of energy, which is used to make things work: The word energy is Greek, and means "the work within." Energy comes in many forms, which can be stored and used in different ways.

The universe was born 15 billion years ago in an incredibly hot ball of energy (left). It began to expand at an astonishing rate, creating matter, and cooling rapidly as it grew. After 10,000 years, atoms appeared and in two billion years began to group together to form stars and galaxies. Whirling gases around the stars condensed into the planets, like Earth and Mars, about five billion years ago.

Matter and energy
Energy and matter seem very different, but they're not: in fact, matter can be turned into energy. This is how the Sun produces its enormous energy, and how nuclear power stations and nuclear bombs work. The physicist Albert Einstein showed that the amount of energy produced is given by the equation $E = mc^2$, where E is energy, m is mass, and c is a very large number – the speed of light.

The energy we need comes from food. Some foods, like sugar or fat, contain more energy than others. In the body, food is digested to release its energy, which then flows through the bloodstream to the muscles. An active person needs more food than somebody who sits down all day.

Aristotle and Galileo

The Greek philosopher Aristotle (384-322 B.C.) was among the first to try to explain energy. He believed that a heavier stone would fall faster than a lighter one – but he never tested it. If he had, he would have found that both fell at the same rate. They did not fall simply because they were heavy, but because they had energy from being lifted. It was the Italian physicist Galileo Galilei (1564-1642) who first began to understand this and to challenge many of Aristotle's incorrect theories.

All machines need a source of energy. Viking longboats were driven by muscle power, which is limited. But the gasoline or diesel engines in a modern mechanical digger are far more powerful. Until the invention in the 19th century of engines that could turn heat from coal or oil into work, the methods by which people altered their environment were very different.

Energy expressions

There are many sayings which use ideas about heat and energy. To "get into hot water" means to get into some kind of difficult situation. To "go full steam ahead" means to do something with all your energy. To "blow hot and cold" is to change your attitude toward something many times. Do you know these sayings: to "be too hot to handle," or to "be firing on all cylinders?"

Tidal waves

The energy of earthquakes under the sea can travel across the oceans as huge waves called tsunamis. The tsunamis, or tidal waves, are barely visible out at sea, but when they reach land they sweep ashore, causing destruction. Systems around the Pacific Ocean warn people to move if a tsunami approaches.

Sound is a form of energy, created by vibrations, such as the sound of a tuning fork. It is transmitted through the air in waves, which travel at about 700 mph (1,126 kph). The vibrations of the air are picked up by our ear drums, which vibrate in time, sending signals to our brains.

"TO BOLDLY GO..."

Space is the final frontier, the explorer's ultimate and infinite goal. Even before exploration of our own planet was complete, people started to investigate other worlds. *Sputnik I*, the first man-made satellite, orbited the Earth in 1957. Three years later the Soviets launched Yuri Gagarin into orbit. In 1966 *Venera 3* crash-landed on Venus. When astronauts Neil Armstrong and Buzz Aldrin (*left*) became the first humans to walk on the moon in 1969, they achieved a thousand-year-old human dream.

Space stations and unmanned voyages of discovery to distant planets followed. At great expense, we are gradually beginning to piece together a picture of our larger environment. The last and greatest age of exploration has begun.

MOON DREAMS. In 165 A.D. Lucian of Samos said moon travelers would find a huge mirror suspended over a well! H.G. Wells's novel *The First Men in the Moon* (1901) described explorers finding small humanoid moon-creatures (*right*).

Fifty-eight years later the Soviet probe *Luna 2* hit the moon. Finally, ten years later, American astronauts confirmed that there were no moon-men or mirrors – only dust and rocks!

LIVING IN SPACE
Space Station Alpha (*above*) will be the first continuously occupied space station, a vital stepping stone toward future crewed missions.

BURNING MONEY
The enormous cost of putting rockets like *Saturn V* (*above*) into space led to the space shuttle. But at $250 million per flight, they are just as expensive.

Comet Chaser
Giotto (right) *was one of many probes that studied Halley's Comet in 1986.*

Saturn V

Giotto

SATURN

URANUS

NEPTUNE

Voyager

Robot Explorers

For centuries, humans had dreamed up strange creatures from Mars, but when the Pathfinder probe (below) *touched down on the surface of the red planet in 1997, no signs of life were found. However, such probes have proved that exploration is possible without the huge risk and expense of putting humans into space.*

Viking

JUPITER

Pioneer

Hello Universe
Pioneer *probes have now left the solar system, carrying a welcome message from humankind into outer space.*

Solar Navigators
Two Voyager *satellites saw volcanoes on Jupiter's moon, Io, storms racing around Neptune, and found new moons around Saturn.*

SMALL, GREEN, AND SLIMY?

Past generations inhabited unknown regions with mermaids, sea monsters, and giants.

They also wondered what strange creatures inhabited the heavens, like this creature from Sirius (the Dog Star), drawn in the 18th century (*right*).

Today, we fill the distant universe with creatures of our own imagination.

Friend or Foe?
If we do meet other intelligent lifeforms, will they be friendly, like E.T. (left)? And will we conquer or wipe them out like so many species and peoples on our own planet?

BEAM ME UP SCOTTY. As a T.V. series and films, *Star Trek* (*below*) remains the most popular modern exploration fantasy. Are the heroics of explorers on the edge of the universe so different from those of their 15th-century counterparts on the edge of the Atlantic?

We have come a long way from our prehistoric ancestors trekking across the Earth, but we still wonder – what is out there – and can we beat it?

BREATHING

Like any living thing, the brain needs oxygen. After a few minutes without oxygen, it would start to fade and die – and we wouldn't want that.

Oxygen is an invisible gas that makes up one-fifth of the air around the body. But it cannot seep through the skull bones, straight into the brain. It must first go down the windpipe into the lungs, then into the blood, and finally to the brain.

Rib breathing muscles

HIC, PUFF, PANT

Sometimes the smooth in-out movements of breathing are interrupted. Hiccups are uncontrolled in-breaths, usually when the main breathing muscle, the diaphragm, gets stretched by a too-full stomach just below it. When the body is very active, its muscles need more oxygen, so the breathing gets faster and we pant.

TUBE-CLEANING
Tiny hairs, cilia, and slimy mucus line the breathing tubes. The mucus traps dust and germs; the cilia wave to sweep it upward, to be swallowed.

SPIES ON YOUR INSIDES

Modern X-ray machines can do wondrous things, especially when helped by computers. In the bronchoscopy, the breathing tubes are lined with a special fluid that shows up on the X-ray picture. A computer can color the picture to make it easier to see the details.

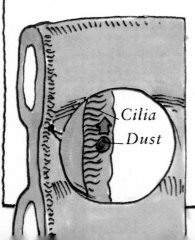

Cilia
Dust

RIBS
These long, springy bones form a protective cage for the delicate heart and lungs. They are hinged so that they move up and down when breathing.

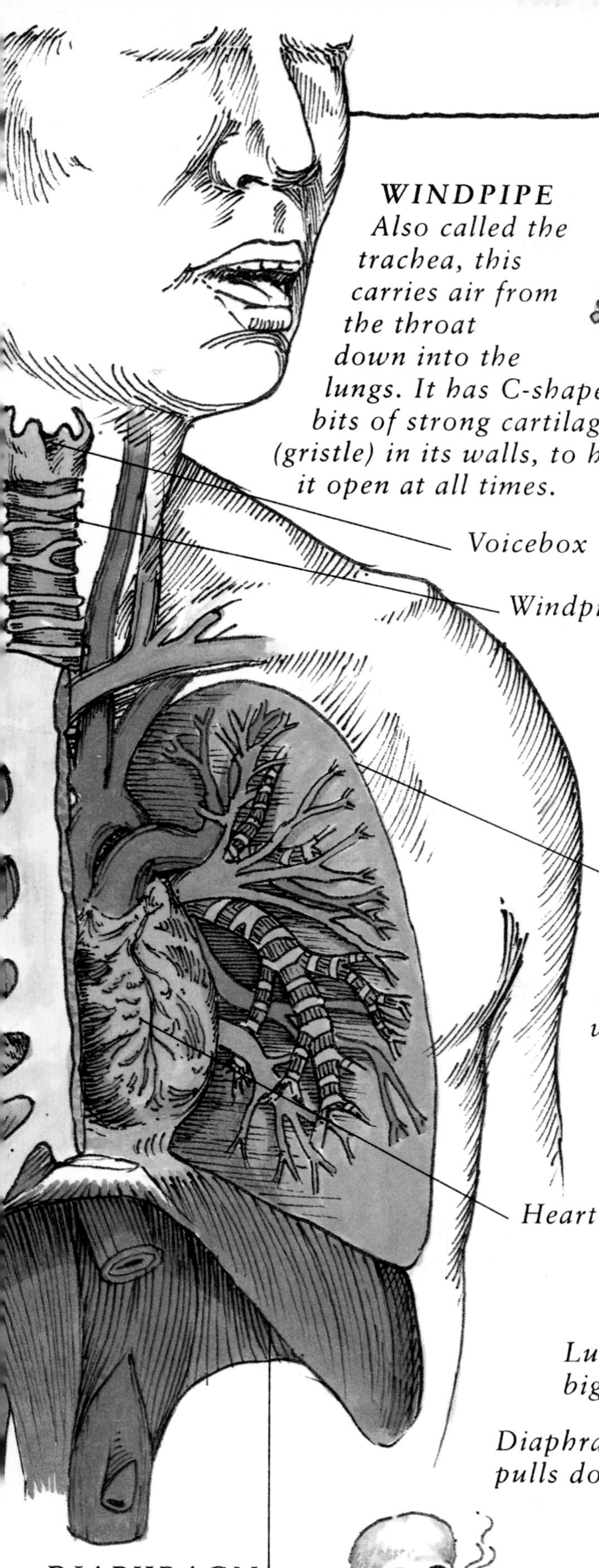

WINDPIPE
Also called the trachea, this carries air from the throat down into the lungs. It has C-shaped bits of strong cartilage (gristle) in its walls, to hold it open at all times.

Voicebox

Windpipe

AIR PIPE
A scuba diver's air hose has strong rings to keep it open, just like the windpipe.

SPARE-PART SURGERY
The windpipe is sometimes affected by disease, especially in people who smoke. Surgeons can replace the affected part with a plastic version that looks like a hairdryer tube.

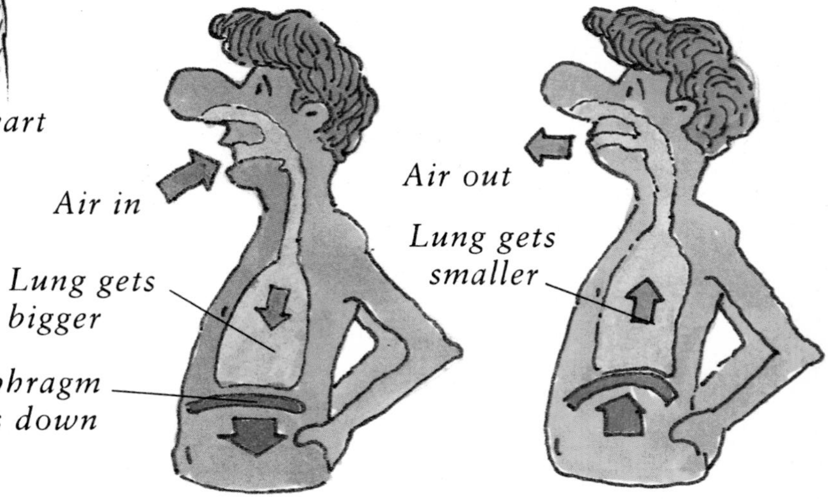

LUNGS
Each of the two lungs has three branching "trees" inside it. These are airways, arteries and veins.

Heart

Air in

Lung gets bigger

Diaphragm pulls down

Air out

Lung gets smaller

DIAPHRAGM
This is a large, domed sheet of muscle under the lungs. With the rib muscles, it powers breathing motions.

SMOKING
The tar, gases, nicotine and other chemicals in tobacco smoke clog and harm the lungs.

IN AND OUT
The brain orders the diaphragm to contract, pulling the bases of the lungs down. This makes the lungs bigger and sucks in air. To breathe out, the diaphragm relaxes.

UFOs and ALIENS

40 million UFO sightings have been recorded since 1947, when the first "flying saucers" were reported. The most common sightings are of glowing balls of light, moving quickly. There have been reports of crash landings by mysterious objects and of alien abductions. In 1948, the U.S. Air Force set up an investigation into UFOs. By 1969, when the project ended, about 12,000 incidents had been recorded. A quarter were caused by natural phenomena or known objects, but the rest remained unidentified. It has yet to be proved that there are intelligent life forms elsewhere in space – or that they are visiting Earth.

Where do UFOs come from?
Most ufologists believe that aliens visit Earth from distant galaxies which human science and technology are not yet advanced enough to find. However, some people have a theory that UFOs come from a hollow area in the center of the Earth and fly into space through a hole at the North Pole!

KIDNAPPED!
People who claim to have been abducted by aliens often describe strange noises, flashing lights, and blackouts. Some say they were examined, losing hair and fingernails, and developing strange marks on their skin.

TIME TRAVELERS
No other planet in our solar system can support life, so UFOs must come from planets orbiting another star like our Sun. It would take thousands of years to reach us from the nearest star. Some people think that aliens can "beam" their ships across space and time.

MYSTERY AT ROSWELL
In 1995, British ufologists unveiled an old film showing U.S. scientists examining the corpse of an alien (below). It has been linked to a reported UFO crash near Roswell, NM, in 1947. The U.S. government claimed the wreckage was a weather balloon, but others said they were hiding a "spy" balloon...or a UFO. The film is now being tested – perhaps the truth will be known at last.

TOP SECRET

CLOSE ENCOUNTERS

UFO sightings are called Close Encounters. *A Close Encounter of the First Kind is seeing a UFO. The Second Kind includes evidence such as landing marks. The Third Kind is when a witness sees or meets alien beings. In 1947, a pilot saw some strange disks in the sky. He told reporters that they looked like "saucers" and the name "flying saucers" caught on.*

FACING THE ALIENS

UFOs appear in many shapes and sizes! On April 24, 1964, a police officer claimed that he saw an egg-shaped craft land and two small, human-like creatures climb out. The aliens saw him, rushed back to the ship, and took off. Scorch marks were later found where the ship had been standing.

Tracking alien beings

Everyone has seen imaginary UFOs in films such as *Close Encounters of the Third Kind* (below), but ufologists investigate UFO sightings by real-life witnesses. They use hi-tech equipment to measure radio waves and magnetic effects which might be caused by UFOs, and track mysterious craft on radar screens. Amateur ufologists watch the skies using telescopes and cameras. If they spot a UFO, they record its position, movement, color, and shape, and send this data to UFO organizations. Permanent observers now keep watch for UFOs around the world.

ENCOUNTER OVER IRAN

On September 9, 1976, a UFO was seen over Iran. Two planes went to investigate, but their controls jammed. The UFO, which was about 160 feet long, seemed to fire at them before speeding away.

RAILROADS

The first steam-powered trains were built in the early 1800s by Richard Trevithick in England and Oliver Evans in the United States. In 1830, George and Robert Stephenson's *Rocket* managed a speed of 20 mph (32 kmh). At first people thought that the force of such speeds would harm passengers! The railroads, however, expanded rapidly over the 19th century and brought far-away places closer.

SIGNALING
In the early days of the railroads, signals were given by flags (*above*).

EARLY RAILROADS
Before steam locomotives, horses were used to pull trucks along short railroad lines linking coal mines to canals and ports (*above*).

"Gladstone" (1880s)

Mallard

ICE

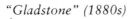

MALLARD

COAST TO COAST
Railroads were vital to the expansion of the United States in the 19th century. Many U.S. steam trains had "cow catchers" on the front (*below*) to reduce the impact of collisions.

CITY RAILROADS

As cities grew in size at the start of the 20th century, new forms of rail transportation were developed. Electric trains are well-suited to urban areas:

They can stop and start quickly; they are quiet and, unlike steam trains, they produce no soot. The world's first public electric railroad was opened in Germany in 1881. To avoid congestion, some urban trains run underground or, like the automated, driverless trains of the Docklands Light Railway in London (*above*), travel overhead.

STATION TO STATION

The British were thrilled with the railroads. By 1900 more than 18,641 miles (30,000 kilometers) of railroad line had been built in Britain alone. The United States now has over 200,000 miles (320,000 kilometers). Cathedral-like railroad stations (*right*) were built in many cities. Waterloo-International (*below*) is a spacious new station in London that services the Eurostar trains that run through the Channel Tunnel to France.

HIGH-SPEED TRAINS (*BELOW LEFT*)
The first steam trains traveled at a horse's pace but, by the end of the 19th century, some trains reached 100mph (160 kmh). The fastest steam train is *Mallard*; it reached 126 mph (202 kmh) in 1938. Modern high-speed trains, like Germany's Intercity Express (ICE) and France's TGV, travel at 186 mph (300 kmh).

ORIENT EXPRESS
The world's most luxurious train, the *Orient Express*, ran from Paris to Istanbul in the early 20th century. It began running again in the 1980s (*above*).

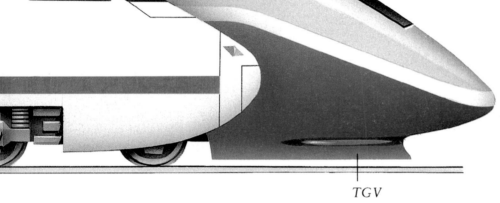

TGV

FREIGHT TRAINS

The development of train travel meant that fresh food could be transported cheaply into cities. Lengthy freight trains are common in the United States and Canada (*below*) where the huge distances make rail freight economical. The longest freight train, which ran in South Africa, had 660 cars, 16 locomotives, and was 4.5 miles (7.3 kilometers) long!

GEORGE STEPHENSON (1781-1848)
Called "the father of the railways," Stephenson was a self-educated engineer whose Rocket (above) won a competition in England for the best locomotive. He helped to prove that railroads could be successful. He also showed how railroad lines could be built almost anywhere.

THE TANK

The battle tank is the main weapon of modern land combat. Its job is to disable or destroy enemy tanks. Every tank is also therefore itself the target of another tank. It may also be attacked by a range of lethal weapons carried by soldiers and aircraft. To survive so that it can do the job it was designed for, it must be able to protect itself from these attacks.

The design of every tank is a combination of three important factors – mobility, protection and firepower. A powerful engine driving a pair of metal tracks gives it mobility. It is protected by a thick covering of heavy armor plate. Firepower may be provided by any of a variety of weapons, but by far the most important is the tank's main gun. This is mounted in a rotating compartment called a turret.

Some tanks are designed for very high speed and mobility. To save weight, they may carry less armor. Others are designed for maximum firepower and protection. The extra weight they have to carry reduces their mobility. This is why tanks come in many different shapes and sizes. Tank designers balance the three basic requirements in different ways.

The structure of the tank has two main components, the turret and the body, or hull. The hull must be large enough to hold the engine, fuel, weapon systems, ammunition and the tank's electronic systems, with enough space left over for the tank's crew of three or four.

A tank's electronic systems include fire control and radio communications. Fire control is a computerized system that helps the gunner to aim the main gun accurately. The tank may also be equipped with specialized instruments

105mm low recoil gun

Wing mirror

Engine dials

Driver's controls

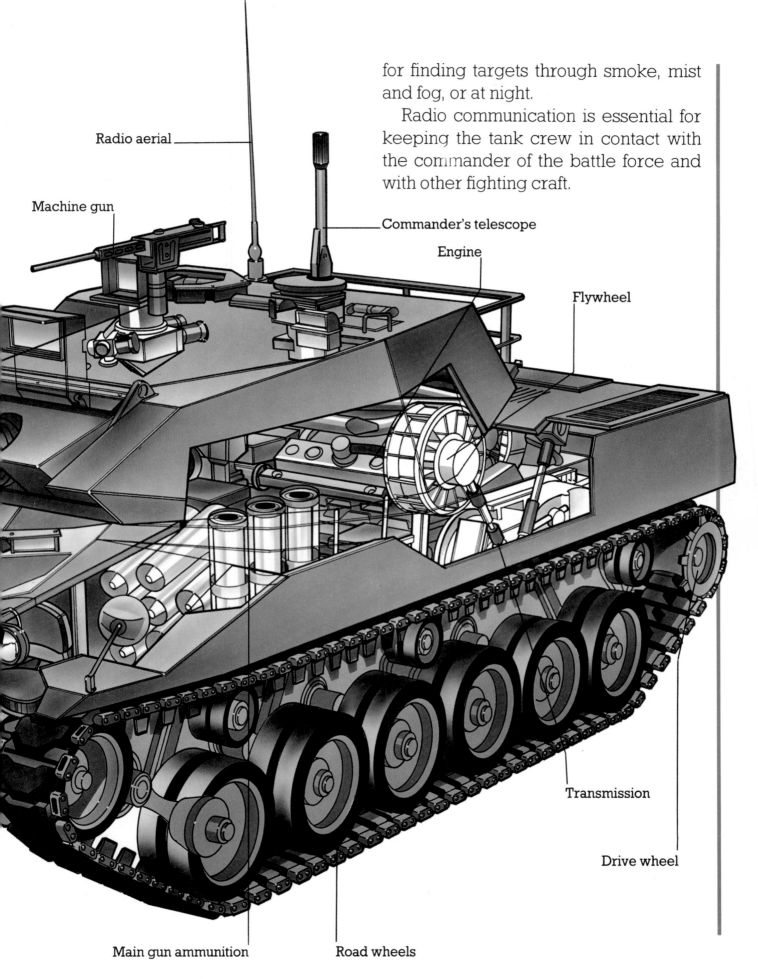

for finding targets through smoke, mist and fog, or at night.

Radio communication is essential for keeping the tank crew in contact with the commander of the battle force and with other fighting craft.

Radio aerial

Machine gun

Commander's telescope

Engine

Flywheel

Transmission

Drive wheel

Main gun ammunition

Road wheels

THE TELEPHONE

"Can I order tickets for Friday by phone, please."
"Did you hear about the great party last week!"
"Emergency, which service do you require?"
"Hello, is that you, Mom?"
A world without telephones would be a difficult
place. We could only talk to people and pass on urgent
messages when they were in front of us. Even modern fax
machines and computer modems rely on telephone lines.

For thousands of years, communication was face-to-face. Only a few simple methods, like smoke signals from a fire, or the beating of drums, could be used to send messages quickly across long distances.

By the 1830s, battery power had arrived. People realized that if they had very long wires, they could send electric signals over great distances, using an on-off switch. The telegraph system was invented, and Samuel Morse came up with his dot-dash code.

The Morse key

Sound to electricity
Soon after the telegraph, inventors dreamed of making the electric signals copy the pattern of someone's voice.

Alexander Graham Bell was a doctor and speech teacher for deaf people. He knew about voices and sounds. In about 1876, he made a simple machine that changed sounds to electrical signals. The signals flashed along a wire, almost a million times faster than sounds went through the air. A similar machine at the other end of the wire changed the signals back into sounds.

EARLY 1920s TELEPHONE

Mouthpiece

Varying current causes vibrations in diaphragm

Sound waves vibrate carbon granules to create a varying current

Earpiece stand

Earpiece

Numbered dial for calling through an automatic exchange

Magnet

BELL'S BOX TELEPHONE (1876)

Alexander Graham Bell

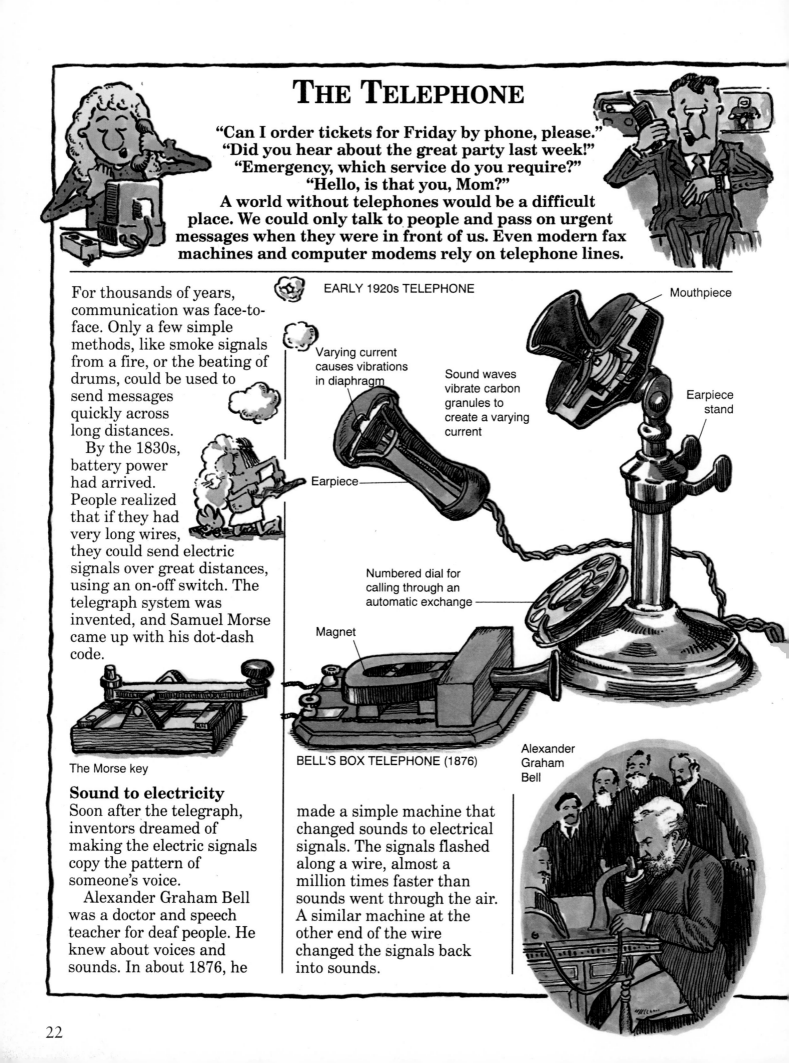

Coast-to-coast phones

In 1877, Bell showed that his machine could send signals almost 20 miles, from Boston to Salem, Massachusetts. Within a few years, telephones were being installed in important buildings and in the homes of rich people.

At first, when you called someone, all the connections were made by hand. Operators worked switches and plugs in the local exchange. The first automatic switches arrived in 1892. By 1915, Americans could phone coast-to-coast. Today, many phone systems use satellite links.

All the same signals

There are probably almost one billion phones in the world. We use them for shopping, passing on messages, doing business, finding out information (such as the sports scores), and simply chatting.

When you talk into a telephone, it converts your voice to electrical signals. The phone lines can carry any similar type of small signals. So they can pass on signals from computers, teletypes, radios, televisions, fax machines, and many other gadgets.

How the telephone works

A telephone has two main parts. These are called the mouthpiece and the earpiece.

In the mouthpiece, sounds make a flat piece of metal vibrate. This squashes and stretches tiny pieces of carbon in a container. Electricity goes through carbon pieces more easily when they are squashed, and less when they are stretched. So sounds are converted to very fast-changing electrical pulses or signals.

The signals go along the wire to the earpiece of the other telephone. They pass through a coil of wire, called an electromagnet. The magnetism produced pulls on a nearby sheet of metal. The strength of the magnetism varies with the fast-changing signals, so the metal moves back and forth very quickly. This makes the sound waves that you hear.

MODERN TELEPHONE — Earpiece

Mouthpiece

- Thin metal sheet
- Electrical signal
- Electromagnet
- Sound waves

Electrical signal

Carbon granules

Soundwaves

Whoops, a bit of trouble

- Alexander Graham Bell first spoke on the phone by accident. A test system was set up in his workroom, when he spilled some acid. He called to his assistant, Thomas Watson, "Mr. Watson, come here, I want you!" In the next room, Watson heard the words over the test system. The first phone call was a plea for help!

- Many phones now have push buttons instead of a circular dial. But people still say, "Dial this number."
- Most phones have a bleeper or buzzer instead of a bell. Yet people still say, "Give me a ring."

AIRPOWER
STRIVING FOR STEALTH

Exotic shapes and sophisticated new electronic systems have created a generation of planes invisible to radar. The experience of surface-to-air missiles in wars in the Middle East and Vietnam convinced strategists that in future aircraft would have to elude detection by radar and attack by heat-seeking missiles. The first practical stealth war planes emerged in the 1991 Gulf War – the U.S. B-2 stealth bomber and the F-117A stealth fighter. Stealth technology means making an aircraft hard to detect in every way: by radar, by sight, by sound, or by the heat of its engines.

The engines of a conventional bomber give off a huge amount of heat in producing their 75 megawatts of power, while a modern infrared detector is sensitive enough to track a lighted cigarette at 30 miles. To make the engines of the B-2 less obvious to the detectors, they are designed to be quiet and cool, and are tucked away in the base of the wings. On the F-117A fighter, known to pilots as the "Wobbly Goblin," the two engines are buried in the thick inner portion of the wings. The fuselage and wing are designed as one because the wing root, where wing and fuselage join, can create a very strong radar echo.

(Below) The Aurora – this reconstruction is what some people think the mystery plane looks like. If it exists, it would be able to fly very high, very fast, and for thousands of miles without refueling on reconnaissance missions over enemy territory.

U.S.A.F

The Northrop B-2 bomber, which first flew in 1989, looks as if it is all wing and no plane. The smooth shape reduces radar echoes by minimizing sharp angles and vertical surfaces. For the same reason, the B-2 has no fin. This would normally make it impossible to fly, so computerized fly-by-wire techniques are used to keep it under control. The B-2 also has complex electronic systems for confusing the enemy, and special paints and materials for a lower radar signature.

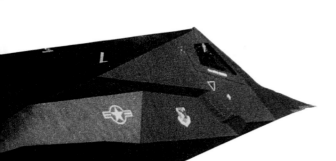

The Lockheed F-117A (above left) does have a tail, but not a vertical one. Its two fins form a V-shape, reducing radar echo. The angles and facets of the plane reflect light and radar in every direction like a cut jewel, confusing defenders. To reduce heat, the two General Electric turbofan engines are deeply buried and not fitted with afterburners, though this does mean a big loss of power.

CODE NAME AURORA
A HIGH-FLYING MYSTERY

Perhaps surprisingly, no order has been placed for a type of aircraft where stealth is most important of all, the high-flying surveillance aircraft. This has not prevented people from speculating that such an aircraft is being built in secret, under the code name Aurora (see far left), and occasional sightings of the aircraft over Scotland have fed the rumors. Although aircraft have been developed in secret before, concealing an advanced aircraft like Aurora for so long would be a major achievement. However, disappointingly, it looks as if Aurora is just a fantasy.

Incoming radar signals bounce off the panels of the F-117A in every direction. This means that there is no single strong echo bouncing back to give the aircraft's position away. It merges into the background.

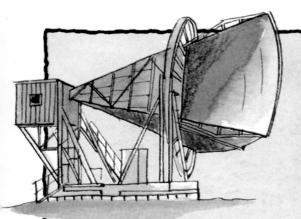

WHAT IF THE UNIVERSE STARTED TO SHRINK?

Echoes in space
Radio telescopes picking up microwaves detect a background "hiss" in space. It shows the temperature of the universe is slightly warmer in some parts than others. These "ripples" are echoes of the Big Bang.

Most experts believe that the universe began as a tiny speck containing all matter, which blew up billions of years ago in a massive explosion, called the Big Bang. It's been getting bigger ever since, as galaxies fly away from one another. This may go on forever, or the universe might reach a certain size, and then maintain a steady state, or it could begin to shrink. All the planets, stars, galaxies, and other matter might squeeze back together to form a tiny speck as the opposite of the Big Bang – the Big Crunch.

What was the Big Bang?

It was the beginning of the universe: the time when all matter began to explode and expand, from a small core full of incredible heat, light, and energy. Was there anything before the Big Bang, like a supreme being? No one knows. There may have been no "before." Space, matter, energy, and even time may have started with the Big Bang.

Strings and clusters

Our galaxy is only one of millions found throughout the universe. Together with a few others, such as the Andromeda galaxy, it forms the *Local Group*, a collection, or cluster of galaxies that circle around space together. The Local Group is, in turn, part of a group of galaxy clusters known as a *supercluster*. These superclusters are linked by massive strings of galaxies that may be up to 300 million light-years long.

How far can we look across our universe?

As telescopes become more powerful, and orbit in space on satellites, they can pick up the faint light and other waves from more distant stars and galaxies. The farthest object currently visible is Quasar (QUASi-stellAR object) 4C41.17. This object is so far away that light reaching us now left the galaxy when the universe was one-fifth its current age.

Happy birthday to you...

The general agreement is that the universe was "born" in the Big Bang about 14 billion years ago. Some believe that it is closer to 17 billion or as low as 10 billion years old. Scientists are still arguing about its exact age. You would need a very big cake for all the birthday candles!

AQUATAIN
THE FLYING BOAT

For years, Western intelligence officials were bewildered by a strange craft, half plane and half boat, which appeared on satellite photographs of the Soviet Union. Now the Cold War is over, its Russian developers have unveiled details of how the craft works. They call it an "ekranoplane" (from the Russian for surface, *îekranoï*) and claim that its high efficiency and low fuel costs come from exploiting something that the early flyers were familiar with – gaining extra lift by flying close to the ground or the sea. Dubbed the "Caspian Sea Monster" by American intelligence experts, the flying boat can skim the surface on an air cushion, something like a hovercraft, but can also soar thousands of feet high to avoid bad weather. Jane's Defence Weekly published the first photographs of this Wing in Ground Effect craft (WIG), and it was immediately recognized as a spectacular piece of technology. The "ekranoplane" can fly over water, land, or ice.

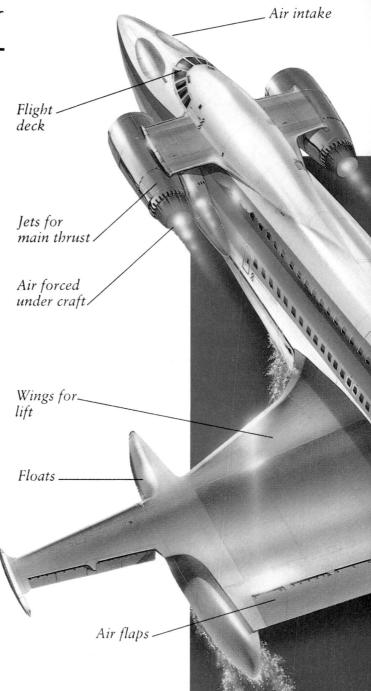

Air intake

Flight deck

Jets for main thrust

Air forced under craft

Wings for lift

Floats

Air flaps

The Airfoil is a new type of boat that is currently being developed in Germany. It uses the same kind of principles as the Aquatain. Short wings on either side of its hull create lift, and help the craft to rise out of the water altogether and fly through the air!

Aquatain would be 240 feet long, and have a wingspan of 180 feet. Carrying 400 passengers, it could be used for short hops such as crossing the Channel, or for longer-haul flights. One problem is licensing the craft, since nobody has yet decided if it is an aircraft, or a boat. Although there are models and videos, the craft has yet to be seen in the West.

A CUSHION OF AIR
HOW THE AQUATAIN FLIES

As designed, Aquatain could fly above water, land, or ice. It uses two sets of engines, one to provide forward propulsion and a second set angled downward to direct thrust under the wings. For takeoff, a set of deep flaps, called screens, are lowered from the back of the wings, trapping the exhaust gases from the second set of engines and creating a region of compressed air, that has the effect of lifting the aircraft away from the water. The forward engines are started, and the craft moves forward, enabling the lift engines to be switched off. The short, broad wings maintain the air cushion on which the craft floats at a height of 45 feet. If necessary, Aquatain can fly much higher, like a conventional aircraft, to avoid storms, but then it is no more economical than any other aircraft. Because few people have ever seen this craft actually fly, some believe that a prototype may have stalled and sunk during a trial flight.

The craft was designed by the Russian Hydrofoil Research Center, and the Soviet Navy has designed several versions. Now the designers want to create a 250-ton craft able to carry 400 passengers at 300 mph, over distances of up to 10,000 miles. This is possible, according to designer Dr. Boris Chubikov, because the craft uses only a fifth as much fuel as a conventional aircraft.

When cruising above the sea, propulsion is provided by two front engines.

Takeoff is achieved by directing the thrust of the lift engines downward, using deep flaps or screens at the back of the wings. This creates a cushion of air.

When the craft is aloft, the main engines propel it forward at up to 350 mph, with the wings maintaining the air cushion underneath.

BLOOD

No matter how frightened the brain is, it never faints at the sight of blood, because blood brings life. Blood carries vital oxygen, energy in the form of blood sugars, nourishing nutrients for growth and repair, the chemical messengers called hormones, and dozens of other essential substances. So smile and be thankful for this red, endlessly flowing river of life.

A heavy ball

BLOOD GROUPS

All blood is not the same. Each body has its own blood group. Doctors found this when they tried transfusing blood (opposite), and often failed. One set of blood groups is A, B, AB, or O. Another is Rh positive or negative, named because it was discovered in rhesus monkeys.

Red blood cell

Platelet

White blood cell

SELF-SEALING SYSTEM

If a blood vessel springs a small leak, it soon seals and mends itself, by forming a sticky scab. Some car radiators do the same (with water).

BLOOD CELLS
Red cells carry oxygen. White cells fight germs. Platelets help blood to clot.

Injury causes leak

Scab seals leak

THREE TUBES FOR BLOOD

Big arteries carry blood from the heart. They divide into capillaries, which are tiny. These join into veins, and return blood to the heart.

Capillary

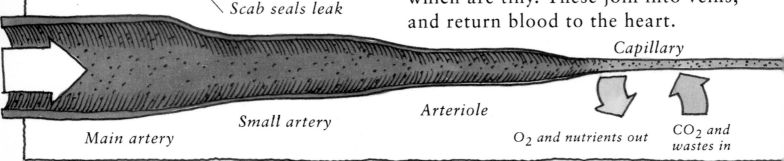

Main artery

Small artery

Arteriole

O₂ and nutrients out

CO₂ and wastes in

CAROTID ARTERY
This tube brings blood to the face, head and brain.

JUGULAR VEIN
It takes head and brain blood back to the heart.

Subclavian artery and vein

Heart

Aorta (main artery)

Vena cava (main vein)

Iliac artery

Iliac vein

Femoral artery

Femoral vein

Saphenous vein

Tibial artery

Pedal arteries and veins

Venule

SPARE-PART SURGERY

Been to the blood bank recently? Most people can safely give, or donate, a small amount of blood. It is treated and put in cold storage. It can be transfused into a person who is injured or ill and needs extra blood.

BODY-WIDE BLOOD

The body's eight or nine pints of blood flow around a network of arteries, capillaries, and veins, called the circulatory system. The regular pumping of the heart keeps blood on the move.

SPIES ON YOUR INSIDES

The angiogram is another type of X ray picture. It displays blood vessels that have been injected with a special chemical, which shows up and reveals any blocks or leaks in the tubes.

Main vein

MEASURING TIME

Before the invention of the clock, people had to rely on nature's timekeepers – the Sun, the Moon and the stars. The daily movement of the Sun across the sky provided the simplest unit, the solar day. The time period of a year was estimated by watching the seasons, and the constancy of the lunar cycle led to the division of each year into months. Traditionally, calendars were controlled by priests. They were devised either by counting days or by following the phases of the Moon. Nowadays, the Gregorian calendar is most common. This was worked out by Pope Gregory XIII in the 1580s.

DAY BY DAY

1. Cut out two circles of cardboard. The largest should be 12in across, the other 11in across. Stick one on top of another and divide into 12 equal pieces to indicate the months.

1

2. Cut out 12 paper circles of 0.5in across. Find out the number of days for each month and write them around the edge of each small circle. Now stick them in order around the large circle. The first day of the month should be nearest the edge as shown.

2

3. Cut another cardboard circle 10.5in across. Cut a hole, radius 0.5in, to correspond with the position of the paper circles. Cover the hole with stiff, transparent plastic. Attach a red arrow marker, as shown.

3

4. Cut out a cardboard circle, radius 0.5in. Carefully make a tiny "window," to view the date through. Position it over the 0.5in hole, and fix it to the plastic with a paper fastener, so it turns.

4

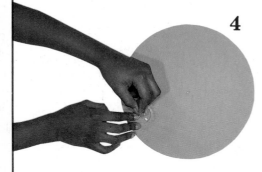

5

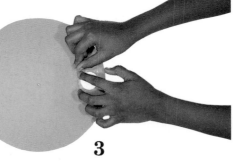

5. Decorate your calendar before joining the separate sections together. Position the smaller circle centrally over the larger circle and join them together with a paper fastener. Rotate your calendar until it is set on the correct day for the current month.

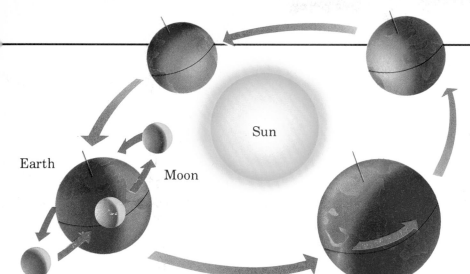

Earth   Sun   Moon

BRIGHT IDEAS

Design a tally system for marking off the days, using symbols instead of numbers. Can you think of a way to group a certain number of tally marks together to make a single tally mark that represents that larger number? We say 7 days equal one whole week. Days and weeks are units of time.

Find out how and when different cultures celebrate New Year. How do their calendars work?

WHY IT WORKS

A calendar is a system of time measurement. Our calendar is based on the movements of the planets. The Earth rotates once every 24 hours, or once a day. The Moon orbits the Earth once every month, and the Earth takes 365 days, or 1 year to orbit the Sun.

6. View the date through the "window." Remember to turn the small wheel daily. Each revolution of the 0.5in circle is equivalent to one month, as represented on the calendar. For each new month rotate the large circle.

6

JUNE JULY MAY APRIL MARCH FEBRUARY JANUARY NOVEMBER OCTOBER SEPTEMBER

WHAT IS AIR?

Air is everywhere on Earth, even inside our own bodies. We cannot see, smell, or hear air but our lives would be very different without it. Air causes changes in the weather, keeps things warm or cool, lets fires burn, and allows sounds to travel. Air consists of a mixture of gases, mainly nitrogen and oxygen, and can be squashed or compressed into small spaces. The air is constantly recycled by nature – so the air we breathe today is the same air that helped plants to grow millions of years ago (above left).

Gases in the air

Most of the air, about 78 percent, is nitrogen gas. About 21 percent is oxygen gas. All living things need oxygen to release energy from their food. The remaining 1 percent of the air consists of gases such as carbon dioxide, argon, neon, helium, krypton, hydrogen, xenon, and ozone.

The carbon dioxide in the air helps to keep the Earth warm. The air also contains dust, and moisture in the form of water vapor.

Argon and other gases 1%

Nitrogen 78%

Oxygen 21%

Humidity

The amount of water vapor in the air is called its humidity. As air cools down, some of the water vapor turns into liquid water, called condensation. This happens because cool air holds less water vapor than warm air. Condensation may cause clouds, fog, or dew to form.

Early morning dew

Pressure

Barometers (see above) measure air pressure, which is caused by the force of gravity pulling the air down towards the Earth's surface. Changes in air pressure signal changes in the weather. High pressure usually indicates fine, settled weather, while low pressure usually means cloudy, rainy weather.

Air studies

Before the 1700s, air was thought to be a pure substance. However, in 1754, Joseph Black discovered carbon dioxide in air. Oxygen was found by Carl Scheele in the early 1770s and by Joseph Priestly (shown right) in 1774. Nitrogen was discovered in 1772 by Daniel Rutherford, but inert gases such as argon were not detected until the 1890s.

Air particles

In a shaft of sunlight, you can often see dust floating in the air. Air always contains many tiny solid particles, from car exhausts, factory smoke, and forest fires. Other sources include pollen from plants and salt from the sea. Polluted air over a large city may contain billions of particles.

Feeding a fire

Fires need the oxygen in the air to burn. People sometimes blow on a fire to give it more oxygen. To put out a fire, they cut off the supply of oxygen by spraying it with water, foam, or carbon dioxide.

Rusty pipes

If iron or steel are exposed to air and moisture, they usually rust and the metal is eaten away. Rust happens when iron joins up with oxygen in the air to form iron oxide. Protective paint can stop oxygen from reaching iron or steel and therefore stops rusting taking place.

Air sayings

The word air is often used in sayings to convey different meanings. For instance, "to walk on air" is to feel elated and for something "to be in the air" means it is uncertain.

Can you find out the meanings of these sayings: to go on the air; to take the air; to give yourself airs; an airy-fairy idea; to be an airhead?

See if you can compose a poem or a song using some common air sayings.

Useful gases

The gases in the air can be collected separately by a process called fractional distillation. Air is made into a liquid by being cooled to very low temperatures. When it warms up, the gases boil off the liquid at different times because they have different boiling points.

Liquid oxygen is used for powering rockets. Oxygen gas (left) is used in breathing apparatus for fire-fighters and sick people.

Nitrogen gas is used to make fertilizers (above right), while nitric acid (a compound of nitrogen and sulfate dissolved in water) is a key ingredient in explosives.

Argon is used to fill the space in most light bulbs as it is an extremely unreactive (or inert) gas.

Green plants use carbon dioxide to make their own food. Carbon dioxide fire extinguishers are used to put out fires in burning liquids and electrical fires. Carbon dioxide also provides the "fizz" in many fizzy drinks.

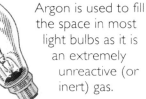

Neon, another colorless, odorless inert gas, is used in fluorescent signs and strip lighting.

Helium is a very light, inert gas used to fill modern airships. It is also used for some types of party balloon.

THE SPACE AGE BEGINS

World War II ended in 1945, but the Cold War, an era of confrontation between the United States and the Soviet Union, began soon after. Both countries developed rockets to display their military might and national pride. Under a brilliant leader, Sergei Korolyev, the Soviets had by 1956 built a giant rocket, the SS-6, capable of carrying a two-ton bomb 4,000 miles. To demonstrate the rocket, Korolyev was ordered to launch a satellite, a small object that would stay in space, and circle around the Earth. On October 4 1957, the satellite Sputnik 1 was launched. The Space Age had begun.

Launchers

Sputnik's SS-6 launcher was big and simple, but very effective. It consisted of a central core, with four strap-on boosters to increase lift off power.

Compared to the Soviet design, the American rockets (below) were lighter and more delicate in construction. They used high-technology fuel tanks instead of the thick-walled steel tanks of the SS-6. Less power meant that American satellites had to be light. This gave a boost to the development of min-iaturized electronic devices such as new transistors, which were soon used in portable radios and other everyday objects.

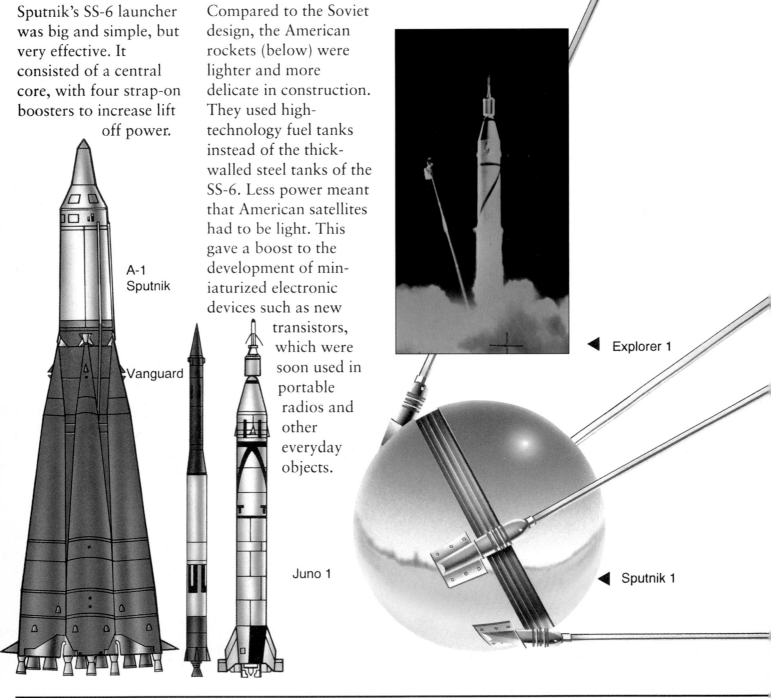

A-1 Sputnik

Vanguard

Juno 1

◀ Explorer 1

◀ Sputnik 1

The response to Sputnik

By late 1957, the United States was ready to match Sputnik with a satellite of its own. But the first launch, of a Vanguard rocket with a tiny 3.4 pound satellite, was a dismal failure. It rose only a few feet before crashing back to the launch pad and exploding in a ball of flame. In desperation, the United States turned to Wernher von Braun, whose satellite project had been in need of money.

American success

Von Braun put together a Jupiter C rocket and a satellite called Explorer. This was launched successfully on January 31, 1958 (see the photo on the opposite page). Explorer 1 was a much smaller satellite than either of the first two Sputniks which preceded it into space. Other early American launches met with less success: a Mercury rocket suffered premature engine cut-off during its launch in 1960 (above left).

In April 1958, the National Aeronautics and Space Administration (NASA) was created, to survey the Moon and put a man into space. It has been a force in world science and politics ever since.

NASA's first office building in Washington D.C.

Sputnik

Sputnik 1 (left) was a simple metal sphere that weighed 184 pounds. Its transmitter emitted a series of beeps. In November 1957, the much larger Sputnik 2 carried a passenger: the dog Laika, who became the first space traveler.

Voices from the sky

A satellite is an object that goes around another. Scientists realized that artificial radio satellites could relay radio, TV, and telephone signals around the Earth. The first was Telstar, (below right) launched by the U.S. in 1962. In 1965, Early Bird became the first geostationary satellite. People could watch the Beatles (below) live on TV beamed from another continent.

The van Allen belts

One of the instruments on Explorer 1 was designed to count and measure electrically charged particles in space. This instrument led to the first space discovery. James van Allen, the scientist responsible, noticed that at certain heights the counter seemed to stop working, and he realized it had been overloaded. The reason was a region in space dense with charged particles - now known as the van Allen belts. These sometimes disrupt radio communications.

CANALS, BRIDGES, AND TUNNELS

All forms of transportation are faced with natural obstacles, and engineers have had to find ways of overcoming them. The ancient Greeks built canals across land as shortcuts for ships and the Romans invented aqueducts – bridges that carried water across valleys. The Romans also coined the word "engineer" from the Latin word *ingenium,* meaning genius. Modern engineering began in the 18th century with the Industrial Revolution. New materials such as iron and steel began to replace stone and brick. The first bridge built entirely of iron (*right*) was finished in 1780 at Ironbridge, England.

EARLY CANALS
The ancient Egyptians (*above*) built canals to irrigate crops in areas that received very little water. Their knowledge of canal-building spread throughout the Middle East to Europe.

towpath

CANAL BOATS
Before boats were fitted with engines, they were towed by horses, who walked alongside on paths known as "towpaths" (*above*). Until the railroads, canals were vital for the transportation of heavy and fragile goods.

CANALS FOR COAL

Canals and rivers carry huge barges carrying thousands of tons of cargo – an inland country such as Switzerland relies on the Rhine River to take goods through Germany to the ocean. Some canals were originally used to transport raw materials like coal, and are now used mostly for canal-boat vacations.

THE SUEZ CANAL
The 114-mile- (184-km-) long Suez Canal (*below*), opened in 1869, cuts through Egypt to link the Mediterranean Sea with the Red Sea. This great feat of engineering took ten years to build. It cut thousands of miles off the trip from India and the Far East to Europe. Today it is a vital route for oil tankers.

LESSEPS

FERDINAND LESSEPS
(1805-1894)
A brilliant French engineer, Lesseps forwarded the idea of the Suez Canal after visiting Egypt. On its completion he was acclaimed as a great hero.

High-speed electric locomotive

Connecting passage

Service tunnel

Running tunnel

THE "CHUNNEL"

The Channel Tunnel (*above*), opened in 1994, makes travel by train or car between Britain and France possible in about 30 minutes. The tunnel is 31 miles (50 kilometers) long – just a few miles shorter than the longest tunnel in the world, at Seikan in Japan.

INVASION THREAT
The first attempt to bore under the English Channel was in 1880 – but progress was halted because of the supposed threat to British defenses (*above left*)!

TOWER BRIDGE
Tower Bridge, London (*left*), is a famous example of a bascule bridge – one in which the bridge tilts upward to let ships pass. Cars drive across when the bridge is down.

FROM LOGS TO SUSPENSION

The first bridges were made of pieces of wood laid across streams. As road travel and railroads increased, bridge building improved. The longest bridges – suspension bridges, such as this one in Normandy, France (*right*) – are held in place by cables supported by towers at each end. San Francisco's Golden Gate Bridge, opened in 1937, has a span of 4,200 feet (1,280 meters).

BUILDING BRIDGES
Long bridges are built in stages (*left*). The weight of each section is supported by intermediate piers crowned with towers, from which parallel cables are suspended.

MAGLEV
THE NEW AGE OF TRAINS

Conventional trains, running on metal wheels on a track, can reach remarkable speeds - over 300 mph, in the case of the French *Train à Grande Vitesse* or TGV. But even greater speeds may be possible for trains that "fly" a few inches above the rails, supported on a cushion of magnetism. Speeds of more than 250 mph have been achieved by the German Transrapid, and the Japanese MLU and HSST prototypes. The Japanese plan is to build a 310-mile maglev track linking Tokyo to Osaka, in one hour.

Sleek and aerodynamic, the MLU00X1 will also be luxurious, with a TV set for every passenger. There will also be a comfortable lounge, and a monitor room fitted with computers, telephones, and other equipment.

The MLU00X1 is the latest version of the maglev train designed by the Japanese company J.R.Tokai. Test vehicles run very smoothly, with only a slight whine from the electric coils, and gentle thuds caused by air pressure in the gaps between sections.

Japan has developed two distinct maglev systems. In the HSST, electromagnets in the wings of the train are wrapped around the guideway and attracted upward toward it, supporting the train in the same way. It reached a speed of 190 mph.

The French Railways, SNCF, does not see that levitation is the future of train travel. Its TGV already links the centers of Paris and Lyons, a distance of 265 miles, in two hours. The TGV Atlantique's record of 320 mph is quicker than any maglev train. When the TGV program began in 1969, SNCF assumed that levitation would be needed to exceed 150 mph, but its first experiments failed, and it turned back to traditional rails.

The MLU00X1 is fitted with eight electromagnets to every coach. At rest, the train sits on wheels, but as it begins to move, the electromagnets induce currents in coils mounted on the floor of the guideway. These currents produce magnetic fields that lift the train off its wheels and support it. Propulsion is provided by coils set in the side of the guideway, which repeatedly reverse polarity, to push and pull the train along.

ATTRACTING AND REPELLING
M O V I N G   M A G N E T S

Two electromagnets may attract one another, or repel one another: it depends on the direction of the current. In the MLU system, the high-power, superconducting magnets, mounted on the train itself, are responsible for inducing opposing currents in the coils on the guideway. When the current is flowing around both sets of coils, they repel each other, lifting the train by between 4 and 8 inches. This generous clearance makes building the guideway easier, but points are difficult to engineer. Because maglev is still being tested, a question mark remains over whether the powerful magnets will have any health effects on passengers, and how high the costs of the track are likely to be.

Guide magnets

Maglev

Magnets

In the past twenty years, magnetically-levitated trains have been developed in Japan, Germany, the U.S., and Britain. One in regular operation in Britain runs along a line just over a mile long linking Birmingham International station to Birmingham airport at a speed of only 15 mph. The German Transrapid system has electromagnets in the wings of the train like the Japanese HSST.

THE MILKY WAY

It is important for an astronomer to know what lies both in and outside our galaxy. Although the stars in the sky seem to be so far away that they are separate from us, the sun and almost all of the stars that we can see actually belong to a single star system called the Milky Way galaxy. The Milky Way contains about 300 trillion stars mingled with clouds of gas and dust. Its shape is similar to a pair of plates placed rim to rim, forming a flattened disk. If we could see it from above, it would look like a vast spiral of light slowly spinning through space. It is so big that even if we could travel at the speed of light, it would take 100,000 years to cross from one side to the other.

WHAT YOU CAN SEE
In some directions, the sky is dense with stars. This is because we are looking through the disk of stars that make up the Milky Way. In other directions, there are few stars against the black backdrop of space. This is the view out of the galaxy, either above it or below the disk of the Milky Way, where there are fewer visible stars.

The arrow shows a spot chosen at random within the galaxy.

The night sky there would appear like this.

The Milky Way forms a hazy band of light across the sky. In the northern sky, it passes through Auriga, Cassiopeia, and Cygnus. In the southern sky, it passes through Vela, Crux, and Sagittarius. It shows up at its best on a cloudless, moonless night, away from city lights.

THE GALAXY'S CENTER

The center of the galaxy is 30,000 light years away in the direction of the constellation Sagittarius. It appears as a dense group of stars in the photo below.

WHERE WE ARE

The solar system (see box below) is situated in one of the spiral arms of the galaxy about two-thirds of the way out from the center. As you can see in this side view, most of the stars lie within the disk shape.

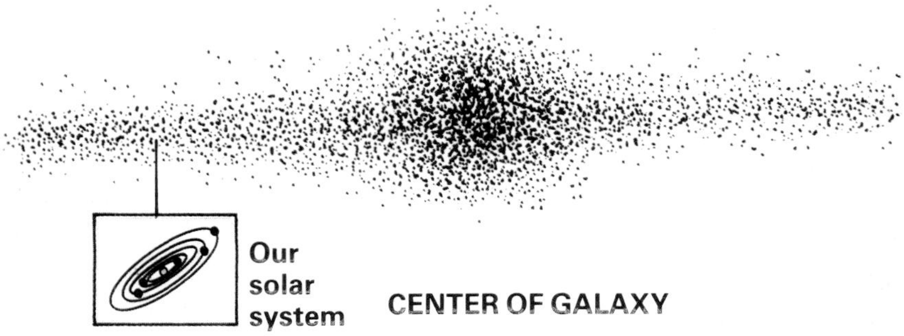

Our solar system

CENTER OF GALAXY

WARPLANES

In less than a century warplanes have developed beyond recognition. The first dogfight between two aircraft took place in October 1914. Since then, air power has been decisive in almost every major war. Today's warplanes carry a wide variety of armaments, including fast-firing cannons, and sophisticated guided missiles and bombs. Some planes, such as the Stealth fighter and the F-15 Eagle, are developed for one purpose only. "Multi-role" warplanes are designed to be able to carry out different functions by modifying a basic aircraft frame.

Fokker Dr-1 Triplane, maximum speed 103 mph (165 km/h)

World War I pilot dressed in warm clothes against the cold

Early warplanes

World War I planes were maneuverable but had few technological aids. The pilot relied on his own flying skills. The top flying ace was German pilot Manfred von Richthofen, known as the "Red Baron," above. His planes – an Albatros, and later a Fokker Triplane – were painted scarlet. Official war records show that he shot down 80 enemy planes, before being killed in 1918.

World War II

Different planes had different roles during World War II. Fighters were small and fast, but could not carry many armaments or fly long distances. Bombers such as the Boeing B-29 Superfortress shown here were bigger, with enough fuel for long flights. But they were slower, and vulnerable to enemy fighters.

F-117A Stealth Fighter

The Gulf Conflict of 1991 was won in the air by planes such as the McDonnell Douglas F-15 Eagle (below). It is a large twin-engined air superiority fighter, specialized to destroy enemy planes in flight.

The coming of Stealth

Ground-based radar systems can detect most enemy planes, unless they are flying very low or between hills. So plane-designers developed so-called stealth technology. The Stealth plane's shape – its curves, edges, and surfaces – are designed to absorb or spread out radar beams, so that they do not reflect back to the receiver. Special paints and surface coatings help this process. Stealth aircraft are designed to be almost invisible on radar, so they can sneak up on the enemy unseen.

Multi-role or specialist?

In recent years, the distinction between small, fast fighters and big, slow bombers has lessened. But there are still specialist warplanes and multi-role craft. The Mig G-25 Foxbat is specialized as an interceptor, designed to tackle enemy bombers at high altitudes. The Panavia Tornado, below, a multi-role aircraft, can carry a variety of weapons at 1,455 mph (2,330 km/h).

Quick escape

If a plane is hit by enemy fire or develops a fault, the pilot has a chance to eject. Pulling a lever opens the cockpit canopy and sets off a small explosive charge, which blasts the seat clear of the plane. A parachute opens, and the pilot sinks to safety.

Inhabiting the
OCEANS

People have always dreamed of living underwater. Many fantastic stories have been written about "human fish" and underwater cities. The first steps have been taken toward making this dream come true. Scientists have developed a membrane that keeps water out but allows oxygen to pass in. It has been tried successfully on rabbits. Could it one day allow humans to breathe freely underwater? Divers are already able to live for some time in "underwater homes," or saturation habitats. These are small and cramped but, one day, larger underwater homes may be created, in which people can live and work for many months or even several years.

BREATHING UNDER THE SEA
In the 1960s, scientist Waldemar Ayres used a special membrane to take oxygen from sea water through artificial "gills." He breathed underwater for over an hour. No one has yet used his system for diving.

WATERY HOUSES
The first "underwater home" for divers was the Conshelf I, *invented by Jacques Cousteau. In 1962, two divers spent a week in it at a depth of 33 feet (10 m).* Sealab *(right) and* Tektite *are habitats in which people have lived for up to 30 days at depths of nearly 660 feet (200 m).*

AN ICY ENVIRONMENT
Even the cold waters of the Arctic and Antarctic are being explored by diving scientists. They have discovered giant sea spiders, anemones, and fish with antifreeze in their blood.

FARMING THE OCEANS

Many countries have sea farms where fish, shrimps, shellfish, (right), and seaweeds are cultivated for food. At present, these are mostly established in shallow water where they can be looked after easily. One day, it may be possible to farm deep-sea fish with diver-farmers living in underwater farmhouses.

How long can people hold their breath underwater?
Most people can hold their breath for about 30 seconds. In December 1994, Francesco "Pipin" Ferreras became the world breath-holding champion when he dived to 420 feet (127 m). He held his breath for 2 minutes, 26 seconds.
WARNING: THIS IS A VERY DANGEROUS THING TO DO AND SHOULD NOT BE TRIED.

TAKING THE PLUNGE

Scuba gear means that almost anyone can learn to dive and explore the deep. But compressed air, used by most divers, is unsafe below about 165 feet (50 m).

Special mixtures of the gases nitrogen or helium and oxygen allow trained divers (left) to reach 330 feet (100 m). Below that, divers use a diving bell as a base, to work down to about 1,320 feet (400 m).

Cities of the future

Many books and films have explored the possibilities of humans living underwater in specially designed towns and cities. Already, a few travel agencies are experimenting with submerged hotels (right) from which guests can watch fish, snorkel, and scuba dive. But as the Earth's population increases, will underwater buildings have more serious uses? Will human colonies be able to live in the depths of the oceans?

ROCKET ENGINE

A rocket is not propelled forwards by the explosive gases rushing from its engine pushing against the surrounding air. For a start, there is no air in space. Three centuries ago the great English scientist Isaac Newton explained it this way: "For every action, there is an equal and opposite reaction." If a shot-putter wearing ice skates throws the shot forward, he moves backward because of the momentum he has created, not because of the shot pushing against the air. Action-and-reaction is the principle of the rocket engine.

A working rocket engine is a "controlled explosion." It burns fuel in an oxidizer (usually oxygen), in a combustion chamber. This creates hot gases under enormous pressure. The gases accelerate out of the back of the chamber. Engineers found that by making a small exit, or throat, from the chamber, the gases accelerate even more, giving extra thrust. They then added a conical nozzle to the throat. This restricts the gases and accelerates them still more, and also helps with guiding the rocket.

Liquid hydrogen tank

The propellant (fuel and oxidizer) tanks are made of specially developed aluminum alloys. They are shaped like giant aerosol cans since they are designed to do the same job – withstand high pressure from within. As the propellants are consumed and the tanks gradually empty, sloshing about of their contents has to be overcome.

THE ENGINE SYSTEMS

The principle of a rocket engine is simple, but there are many practical problems. Engineers have designed various systems to overcome these. In the Space Shuttle main engine, oxygen oxidizer and hydrogen fuel are first pressurized, mixed and preburned, to form hot gases. These gases are then introduced together in an exact mixture in the combustion chamber. The ultracold fuel circulates in a heat-exchanger, to warm itself before preburning and to cool the chamber and nozzle.

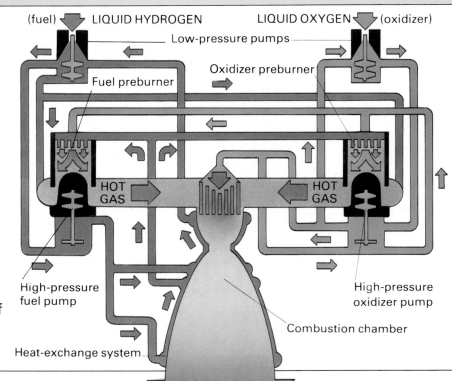

(fuel) LIQUID HYDROGEN   LIQUID OXYGEN (oxidizer)

Low-pressure pumps

Fuel preburner

Oxidizer preburner

HOT GAS   HOT GAS

High-pressure fuel pump

High-pressure oxidizer pump

Combustion chamber

Heat-exchange system

Apollo, the United States' moon program, was launched by the Saturn V rocket. This had a third stage fueled by liquid hydrogen and oxygen. These liquids were fed at high pressure and carefully-controlled rates into the combustion chamber.

Liquid oxygen tank

In this rocket stage, the liquid oxygen tank is contained within the liquid hydrogen tank. The design saves space and weight. Although the liquid oxygen tank is smaller, the weight of its contents is greater than that of liquid hydrogen. A specially adapted Saturn V third stage formed the orbiting space laboratory Skylab in 1973.

Fuel line

Gimbal engine mountings

Oxidizer preburner

Fuel preburner

Heat exchanger

Fuel pump

The engine

The third stage of the Saturn V was powered by one Rocketdyne J-2 engine. This was shielded by a "skirt" until about eight minutes after lift-off, when the second stage separated and fell away. Then the J-2 ignited and burned for about three minutes to take the vehicle into "parking orbit" around the Earth. Several orbits later it fired again for six minutes, to boost the vehicle free of Earth's gravity and towards the 'mission'

SOLID-FUEL ROCKET

The solid-fuel rocket engine does not burn gunpowder, like a toy firework, but a specially mixed propellant. But once ignited it cannot be turned off. It is used mainly as a booster, strapped to the main engine.

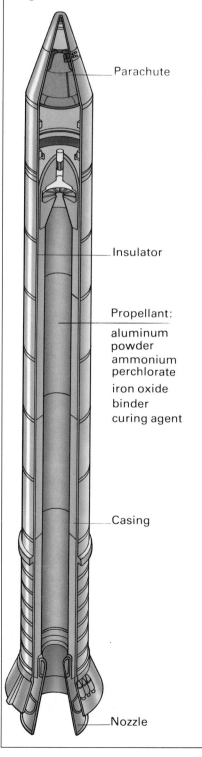

Parachute

Insulator

Propellant:
aluminum
powder
ammonium
perchlorate
iron oxide
binder
curing agent

Casing

Nozzle

49

THE TELESCOPE AND MICROSCOPE

Do you believe that the Moon is made of green cheese, and that germs don't exist? If you had a telescope and a microscope, you could see for yourself. Telescopes look at outer space, to reveal the mysteries of the universe. Microscopes look into inner space, to show us what our own bodies are made of.

Many inventors and scientists fooled around with curved pieces of glass, called lenses. From the 1200s, lenses of various strengths were used in eyeglasses. Two lenses, specially shaped and put near to each other, make distant things look bigger and nearer. This was probably discovered by Hans Lippershey in Holland, in 1608.

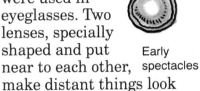

Early spectacles

Within a year, the famous Galileo heard about the new invention, and made his own versions. They magnified up to 30 times. He scanned the dark skies and discovered mountains on the Moon, spots on the Sun, and moons going around Jupiter. He was the first real telescope-user.

Galileo

Types of telescopes

Christian Huygens, who worked on pendulum clocks, also invented better telescopes. In about 1757 a British optician, John Dollond, sandwiched two lenses closely together. These compound lenses gave better, clearer images.

In 1668 the great Isaac Newton designed a telescope with a curved mirror in place of one lens. This was called a reflector. Today, the biggest optical telescopes are reflectors. They look deep into space, to tell us about our Moon and Sun, the planets and stars, and the beginning of the universe.

REFRACTING LENS TELESCOPE

Isaac Newton

Telescope focused by moving extensions

Extensions

Observatory

GALILEO'S TELESCOPE

Eyepiece lens

REFRACTING LENS TELESCOPE

Eyepiece lens

Convex lens

Mirror

NEWTONIAN REFLECTOR

Eyepiece Mirror

THE TELESCOPE AND MICROSCOPE

Seeing the invisible

Around 1590, Dutch lens-maker Zacharias Janssen also put two lenses near to each other. He noticed that a tiny thing at one end looked much bigger from the other end – provided the lenses were the right distance apart.

Scientists showed an interest. They realized that a whole tiny world was waiting to be discovered. In 1655, Robert Hooke first used the word "cell" for a microscopic part of a living thing, in his book Micrographia, "Small Drawings." Hooke showed that tiny creatures such as ants had a heart, stomach and other body parts – just like bigger animals, but smaller!

LEEUWENHOEK'S MICROSCOPE

Lens

Object

Draper turned lens-maker

By the 1680s Anton van Leeuwenhoek, in Holland, made fascinating discoveries through the microscope. His homemade microscopes could magnify over 250 times. Through them he saw new wonders such as red blood cells, tiny one-celled creatures like amoebas, the eggs of fleas, insect eyes, and stacks of cells in the thinnest leaf.

Anton van Leeuwenhoek

After Leeuwenhoek, many people looked through microscopes. The science of microbiology began. Soon people were looking at germs, and working out how they invaded the body and caused diseases.

Microscopes became more powerful, with strong lights and changeable lenses that could magnify over one thousand times. Today, medicine and biology would be lost without microscopes.

MODERN MICROSCOPE

Eyepiece

Lens

Variety of lenses

Mirror

Light path

How they work

In a microscope, light waves from the tiny and nearby object are bent inwards by the convex (bulging) lens. They are bent again at the second lens, the eyepiece, so that you see a clear and enlarged view.

In a telescope, much the same happens. But the light waves come from far, far away, so they are parallel when they reach the telescope.

EARLY MICROSCOPE

Eyepiece

Metal body

Focusing screw

Object

Wow! Big, small, and far-out

• The telescope with the biggest one-piece lens is at the Yerkes Observatory, in the USA. The lens is 40 inches across.

• Electron microscopes use electron beams instead of light rays. They magnify over one million times.

• Radio telescopes detect not light rays, but radio waves from stars, quasars and pulsars. They can "see" billions of miles, to the far side of the Universe.

Radio telescopes

ELECTROLYSIS

Electrolysis is a process in which an electric current is passed through a liquid, causing a chemical reaction to take place. The liquid used is called the electrolyte. The wires or plates where the current enters or leaves the liquid are called electrodes. The electrolysis of metallic solutions is useful in putting metal coatings on objects. If you have a look at some car bumpers, you will notice that they may have a nice, smooth, metallic appearance. This is because they are coated with a metal called nickel, in a process called electroplating. This helps to stop the metal underneath from rusting. The same method is used to coat cutlery with silver. This is called silverplating. Michael Faraday discovered the first law of electrolysis. The process is also used to purify metals like aluminum.

COPPER PLATING

1. For this project you will need a glass jar, a copper coin, a paper clip, two batteries, insulated wire, and water. Pour the water into the jar. Place the batteries together with unlike terminals adjacent. Connect wires to the terminals. Attach the copper to the wire from the positive terminal of the battery. The paper clip must be attached to the wire from the negative terminal. Use modeling clay. Do not allow the metal objects to touch in the solution. You could even tape each wire to the side of the jar so that they are suspended.

2. Observe closely what happens. Can you see bubbles? Leave them for a few minutes, then remove. Observe any color changes. Replace them for a while. Are there any further changes?

WHY IT WORKS

The copper coin is connected to the positive terminal of the battery – the current enters here. The other, the paper clip, is joined to the negative terminal – the current leaves here. As the current flows through the water from the positive electrode (anode) to the negative electrode (cathode), the copper is carried from the coin to the clip.

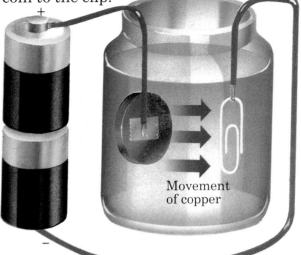

Movement of copper

BRIGHT IDEAS

Repeat the project using salt dissolved in vinegar instead of the water. What difference do you notice – if any? What do you observe about the appearance of the paper clip? Maybe your school has scales that can weigh very small objects? If the coin and the paper clip are weighed before immersion in the liquid and their weight recorded, you can check whether electroplating has really taken place. After carrying out the project weigh them both again. Now replace the battery with a more powerful one, or add a second battery into a parallel circuit, to increase the "push" of the current passing through the liquid. (Remember to stop your experiments if the batteries heat up.) Weigh the coin and paper clip a second time. If the weight of the paper clip has increased further, then you have proved the first law of electrolysis – the size of the charge passed through the liquid determines the amount of copper freed.

2

HOT-AIR FLIGHT

Have you seen a hot-air balloon? More than 200 years ago, two French brothers, Joseph and Jacques Montgolfier, discovered that rising hot air could be captured and used for flight. They made a huge balloon from linen and paper and built a fire underneath it. The balloon trapped the hot air and smoke rising from the fire and lifted the two men into the air. As the air cooled, the balloon floated back down to the ground. Since that first flight, people have used hot-air balloons for pleasure, for racing, and even for warfare. You can make your own hot-air balloon.

BALLOON LIFT-OFF

1

3

1 To make a balloon that traps hot air to fly, you need four large sheets of tissue paper. Fold each sheet in half and lightly copy this shape on to one using a pencil. When you are happy with the outline cut out your first "panel."

3 Unfold your first panel and spread glue on the edge of one half. Stick the second panel on top and press down. Repeat with the next panel until all four panels are joined into a balloon.

4 Make a small "passenger basket" from a piece of folded oak tag. Attach the basket to the open end of the balloon with four lengths of thread.

2 Use the first panel to help you mark out the next three. Cut them out and trim them carefully to make sure they are all the same size.

2

4

5 Take the balloon outside for your first flight. Blow up the balloon with hot air from a hair dryer and watch it lift off.

5

Lift

Balloon

Heated air

Gravity

WHY IT WORKS

Your hot air balloon rises because it contains air that is warmer – and therefore lighter – than the surrounding air. (Air, indeed all gases, expand when heated. They become lighter because the same amount of gas takes up more space.) Hot air from the drier enters the bottom of the balloon and rises inside to the top, causing the balloon to lift off. The colder the air around it, the faster the hot air will rise. A hot air balloon has no power to move along – it needs wind to help it.

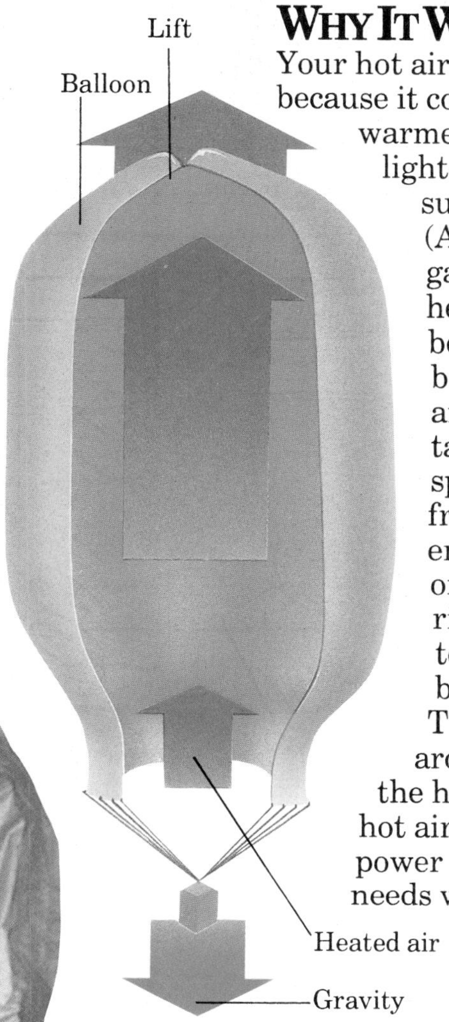

BRIGHT IDEAS

See if your hot-air balloon works better in a hot room or a cold room. (See why it works, above.)

Make some modeling clay passengers for your basket. Notice whether the balloon needs more hot air for lifting power.

Will a larger hot-air balloon rise even better? Build one and find out.

Watch the smoke rising above a campfire. Do you see how the hot air carries it up? As the air cools the smoke stops rising as fast. Notice what happens then. Does the smoke scatter in the wind?

P O W E R

FROM THE SKIES

Clouds and nightfall limit the amount of sunlight reaching the earth. In space, there are no such limits; a solar cell in space would receive 10-15 times as much sunlight as one on earth.

To take advantage of this, American engineer, Peter Glaser, has proposed the concept of a solar power satellite (right). A large number of solar cells in geostationary orbit (orbiting above the same point on the earth's surface) would generate electricity from sunlight. The electricity would be converted into microwaves, beamed to a collector on the earth's surface, and converted back into electricity. In the United States, studies by the space agency NASA, show that such a system could, in theory, supply enormous amounts of electricity day and night.

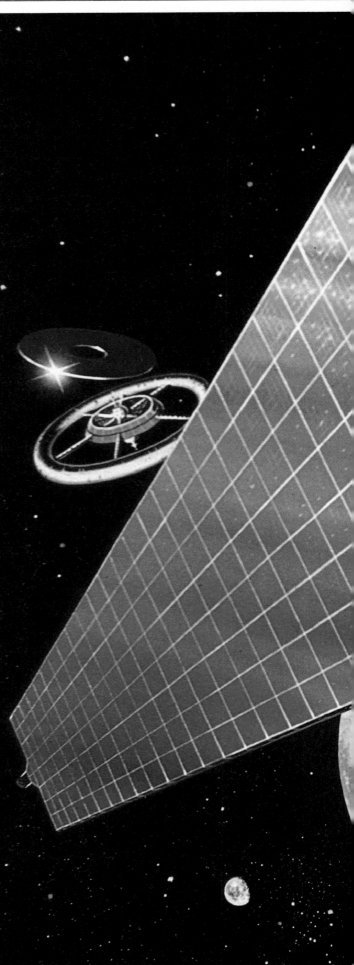

The satellite would be built in space, and placed in orbit 23,000 miles above the earth. In this orbit, the earth would block the sun from the satellite only one percent of the year.

The satellite would consist of solar cells to collect the sunlight, and a transmitter to send it to earth. An alternative would be to place the whole system on the moon.

Reliable methods exist for converting electricity into microwaves; people use them daily in microwave ovens. Microwaves travel through space and easily penetrate the atmosphere. Precise aiming of the beam is important to ensure that it does not stray over populated areas. The collectors (above) could be built in remote areas, or float in shallow water.

SPACE POWER
BUILDING THE SATELLITE

All the technology needed to build the solar power satellite exists, but the scale of the project is enormous. In space, millions of solar cells would have to be assembled into huge collectors. The equipment to transmit the microwaves or laser beam would have to be at least half a mile across. On earth, a collector for the incoming beam would have to be over 36 square miles in size. This would have to be built in a remote place or on a platform over an expanse of water. It would be capable of generating electricity for two million households. The possible effects on health of exposure to the microwave beam need careful investigation, but exposure should be no greater than from radar sets or industrial processes, where no ill effects have been seen. Solar power satellites might be hit by meteorites, and would be difficult to protect in a high-tech war. But they would be pollution free, inexhaustible, and capable of almost indefinite expansion.

Object

Lens

Cornea

Vitreous humor

Retina

Image

Optic nerve

Light rays

RETINA

FOCUSING

Rays of light reflect off objects and pass through the clear dome of the cornea at the front of the eye (above). They then pass through the pupil, the dark hole at the iris' center. The light rays are then focused by the lens, so that they shine through the vitreous humor and form a clear, but upside-down image on the retina. This picture is turned into nerve signals. These pass down the optic nerve to the brain where the image is turned upright.

RETINAL RODS AND CONES

The retina contains up to 126 million microscopic light-sensitive cells, called rods and cones (above). Each one changes light energy into nerve signals. The 125-plus million rods detect low levels of light, but cannot distinguish colors. Cones see colors and details, but they only work in bright light.

IRIS

LENS

Cornea

Sclera

PUPIL

INSIDE THE EYE

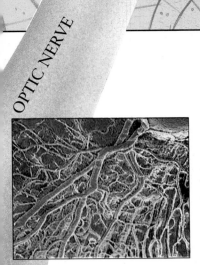

EACH EYEBALL is covered with a tough outer sheath called the sclera, except for the colored part at the front. This colored circle on your eye is called the iris. At its center is the black pupil. The iris changes size to let more or less light through the pupil and into the eye.

Most of the eye's interior is filled with a clear jelly, the vitreous humor. The part that detects light, the retina, lines the rear of the eyeball. This very delicate layer provides the detailed moving pictures of the world that you experience in your brain.

BLOOD VESSELS
The back of the eye is covered by branches of blood vessels (above), which supply the retina with nutrients.

Cataracts affect millions of people around the world and are a major cause of blindness in less-developed countries. This misting or haziness in the lens (left) obscures sight. As it worsens it can cause total loss of vision. Surgery to remove the misty portion or the whole lens, and insert a plastic artificial version (above) or implant, can drastically improve eyesight.

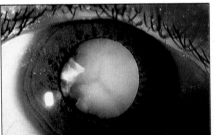

BLOOD VESSELS

Tear duct

FAR SIGHT
The eye is too small compared with the power of its lens, so the image cannot be focused (right). Convex lenses will focus the image on the retina.

SHORT SIGHT
The eye is too large compared with the focusing power of its lens. Concave lenses diverge the light rays before they reach the eye's own lens.

TEARS
The tear gland makes salty tear fluid. This flows onto the eyeball, where the eyelids wipe it across the surface. It helps remove dust and germs. When we cry (below), we produce a lot of tears.

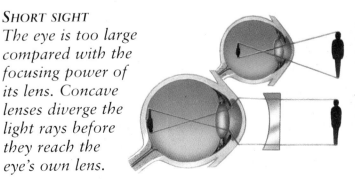

MISSILES AND ROCKETS

BOOMERANGS DON'T ALWAYS COME BACK!
The traditional hunting boomerang of the Australian Aborigines (*above*) was cleverly designed to return to the thrower if it missed the target. However, the heavier war boomerang was designed to fly straight, and did not return to the thrower.

One of the first times a missile was used in war is recorded in the Bible, when David killed Goliath the giant using a slingshot, or catapult. Such weapons were originally developed for hunting, but their military usefulness soon became apparent. All early missile-throwing systems used the power of the human body (*above*), but the Chinese discovery of gunpowder made possible the development of rockets, which had greater power and far greater range.

Medieval European lance

Japanese yari

Roman pilum

African spear

Indian lance

Maori spear (from New Zealand)

AIR POWER
The native peoples of the Amazon region of South America use long blowpipes to hunt animals in the jungle (*right*). A poisoned dart is blown along the tube, and in skilled hands it can be very accurate.

THRUSTING AND THROWING

Spears can be thrust or thrown at an enemy, and there are many variations (*left*). The Roman pilum, for example, had a long iron point that was designed to pierce enemy shields and break off from the wooden shaft, so it could not be thrown back. This Australian aboriginal device (*right*) is attached to the end of a spear to give the user greater force when throwing.

SIEGE CATAPULTS

Invented by the ancient Romans, catapults (*left*) were also used during the Middle Ages to batter down the walls of castles. The throwing arm was attached to a thick cord and winched backward. A large stone was placed in the cup and the throwing arm was then released, hurling the rock forward.

SIR WILLIAM CONGREVE

The British general William Congreve developed rockets during the Napoleonic Wars (1793-1815). Rockets were used against the French at the Battle of Leipzig in 1813 and, although they had a range of 0.8 miles (1.3 km), they were not very accurate.

CHINESE CRACKERS!

The first time a rocket was ever used in war is thought to have been in 1232, when Chinese troops fired gunpowder rockets at a Tartar army (from Russia). These rockets would have been like our modern firework rockets, and they were intended to frighten the enemy's horses.

WORLD WAR II ADVANCES

Rockets became effective weapons during World War II. Germany fired high-explosive warheads with a range of 4.5 miles (7 kilometers) from a multi-launch system called a Nebelwerfer. The United States and

England used a hand-held antitank missile launcher (*left*) nicknamed the "bazooka" after the comic-strip character Bazooka Joe!

UP AND AWAY!

The Stinger surface-to-air missile (*right*) makes it possible for a single infantryman to shoot down low-flying jet aircraft. Developed in the 1980s for the U.S. Army, the Stinger has a range of 3 miles (5 kilometers) and can hit aircraft flying up to 3 miles (5 kilometers) high. After the soldier has pulled the trigger, the missile is guided to the target by an infrared homing device, which locks onto the heat produced by the aircraft's engine.

SOUND WAVES

Sound waves are similar to light waves in some ways. Like a beam of light can be reflected from a mirror, so a sound can be reflected from a surface like a wall. If you shout loudly in your school hall, the sound waves travel to the wall and are bounced back, reaching your ears a split second later – this is an echo. Bats make use of echoes when finding their way, or hunting. They give out a very high pitched sound that bounces back from objects or insects, telling them how far away things are. We can also use this method to find objects that we cannot see. Sonar uses sound to locate objects at the bottom of the sea, such as shipwrecks and shoals of fish.

FIRING WAVES

1. This cannon will send a narrow "beam" of sound waves. Begin by making a pair of wheels for your cannon using circles of cardboard, paper plates, thread spools, and a wooden stick.

1

2. Make a large tube of stiff cardboard for the cannon itself, 18-20 inches in diameter, and four feet long. Make the back of the cannon by covering a circle of cardboard with plastic wrap. Attach **2** with tape.

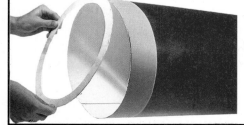

3

3. The front of the cannon is a disk of stiff cardboard with a 1 in hole in the center. You could decorate this with a disk of colored paper.

4

4. Tape the ends of the cannon firmly with double sided tape – this will enable you to fix the ends onto the tube and not into it.

5. Fix the tube to the wheels with tape, and weight the back end of the cannon so that it does not tip forward. Aim the cannon at a wall and tap the plastic wrap quite firmly – from a distance you should get an echo. Make a curtain from 0.5 in strips of foil. Fire your cannon at the curtain. You should see the sound waves making the foil vibrate.

5

WHY IT WORKS

When you strike the piece of plastic wrap (diaphragm) on the cannon, the vibrations lead to sound waves being formed. The waves travel outward from the diaphragm, making the air particles around move back and forth in the same direction. When the waves leaving the front of the cannon meet a solid object, some of them are reflected while some continue traveling through the object, making it move slightly like the air particles. A bat uses sound to find objects in the dark. It produces sounds and then listens for the echoes to be reflected. This is called echolocation.

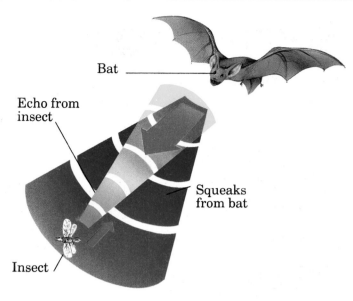

Bat

Echo from insect

Squeaks from bat

Insect

BRIGHT IDEAS

☀ Which surfaces are best for echoing? Can you make an echo in a bathroom, a kitchen, a cafeteria, a hall, a subway? Try other places, too. Do you think hard surfaces are better than soft surfaces for giving echoes? Which sounds echo the best?

☀ Try shouting, knocking two stones together, banging two blocks of wood together, whistling, and talking. Do short, sharp sounds echo better than long soft ones?

☀ The cannon channels the sound waves in one direction. Make a megaphone to channel your voice. Make a narrow cone, and then a wide cone. Shout to a friend through each cone. Shout again, aiming the megaphone 30 feet to their side. What happens? Which cone makes your voice sound louder, and which can you hear best from the side? Can you figure out why?

PLOTTING THE STARS

Astronomy is the study of the stars, planets and other objects in the universe. For centuries, astronomers have striven to learn more about our Universe. Through observation and careful measurement, using scientific tools like the telescope, we now know that the Sun is the center of our solar system. Accurate measurement of star distance is a science developed over the centuries by astronomers like Tycho Brahe (1546-1601). Centuries ago, sailors calculated time at night by observing the movements of star clusters near the fixed Pole Star. Watches aboard ship were timed from the position of these constellations in the sky.

STAR TIME

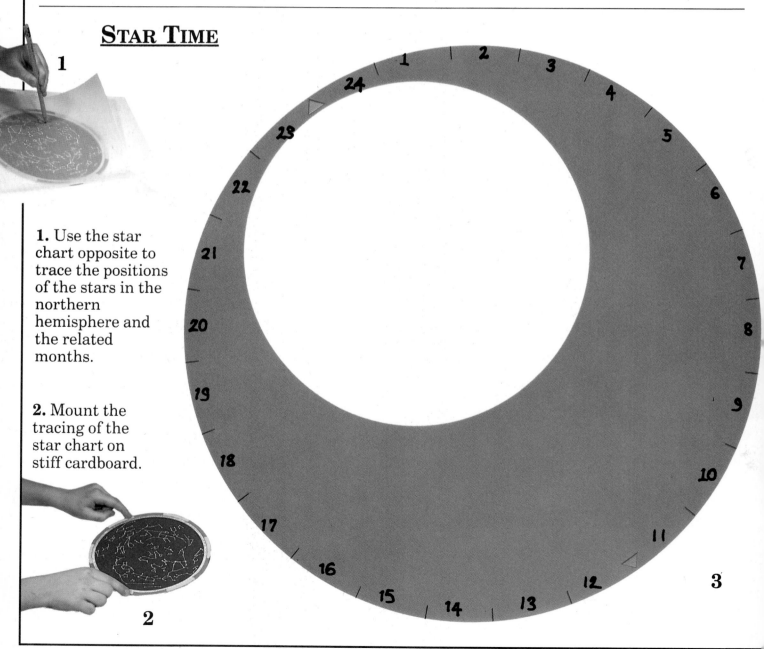

1. Use the star chart opposite to trace the positions of the stars in the northern hemisphere and the related months.

2. Mount the tracing of the star chart on stiff cardboard.

The star chart wheel is labeled around its edge with the months (clockwise from top): NOVEMBER, DECEMBER, JANUARY, FEBRUARY, MARCH, APRIL, MAY, JUNE, JULY, AUGUST, SEPTEMBER, OCTOBER.

3. Trace the shape on the opposite page onto cardboard. Mark the 24 hours of the day, starting with Noon at the bottom.

4. Proceed counterclockwise covering the "window" of the planisphere with transparent plastic. Pin it to the starchart through the center.

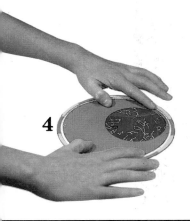

4

5

5. On a starry night, rotate the planisphere until the time of day, marked on the edge, is lined up with the appropriate month at the bottom of the star chart. Compare what you see with the stars in the sky.

WHY IT WORKS

As the Earth revolves around the Sun, different constellations of stars appear and disappear in a continuous cycle that can be observed. This planisphere shows where these groups of stars can be seen at any given time of the year, in the northern hemisphere. By matching up the time of day with the date, you can view the stars that should be visible in the night sky through the cut out "window." The planisphere should be held up and viewed from underneath. The stars visible through the window should match those in the sky. The Sun is a star. It is the only star close enough to look like a ball. The other billions of stars are so far away, they appear to be pinpoints of light.

WHAT ARE ELEMENTS?

Everything in the world is made up of elements. An element is a simple chemical substance which itself is made of tiny particles called atoms. Diamonds (right) are made from the element carbon. There are 105 elements known to man. Some, like iron or copper, were used in ancient times, while others have been made by modern scientists. Many elements can be combined to make new substances – copper and tin can be melted together to make an alloy (mixture) called bronze. The periodic table arranges all the known elements in a special way.

PERIODIC TABLE

1. Each square contains the symbol for a different element. The symbol comes from the common name or Latin name of the element; C stands for carbon and Al for aluminum. The Latin for iron is ferrum, so the symbol is Fe.

H								
Li	Be							
Na	Mg							
K	Ca	SC	Ti	V	Cr	Mn	Fe	C
Rb	Sr	Y	Zr	Nb	Mo	Tc	Ru	R
Cs	Ba	La	Hf	Ta	W	Re	Os	I
Fr	Ra	Ac						

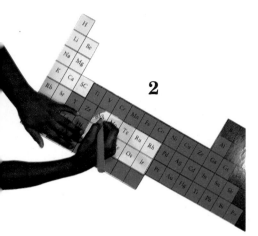

2

3. Now you can begin to search for some things which contain common elements in your own home or school. On the periodic table below, you can see some Na which stands for sodium (in salt), Mg – magnesium (in powdered milk of magnesia), Cr – chromium (used to coat car bumpers and keys), Fe – iron, Cu – copper, Zn – zinc (in a battery) Ag – silver, Sn – tin, S – sulphur (in matches), Pb – lead, Al – aluminum and Au – gold.

3

2. To construct your own periodic table you need to copy the grid below carefully. The table is always set out just as you can see it here and the position of the boxes may not be changed. To find out what all the symbols mean you could visit a library and look for more information.

						He
B	C	N	O	F	Ne
Al	Si	P	S	Cl	Ar
Ni	Cu	Zn	Ga	Ge	As	Se	Br	Kr
Pd	Ag	Cd	In	Sn	Sb	Te	I	Xe
Pt	Au	Hg	Tl	Pb	Bi	Po	At	Rn

Now try to find some... C – carbon (soot, pencil lead, diamonds and charcoal are all forms of carbon), Hg – mercury (the silver stuff in thermometers), Pt – platinum (a metal used in jewelry making) and Cl – chlorine (added to drinking water to kill bacteria).

WHAT IF THERE WERE NO SPACECRAFT?

How would astronauts get back to Earth?

An astronaut could survive in a spacesuit for a short time. But coming back into Earth's atmosphere creates lots of heat, as an object pushes through the ever-thickening air molecules. A heat shield might help a rear-first re-entry!

Space exploration would be much less exciting without spacecraft that carry people. It began with the Space Race in the 1950s and 1960s. The United States and Russia raced to launch the first satellite, the first spaceman and woman, and the first Moon visit. The satellite Sputnik 1, launched in 1957, was the first man-made object in space, and the Russian Yuri Gagarin was the first man in space. But the United States was first to land on the Moon in 1969 with the spacecraft Apollo 11 carrying Neil Armstrong and Edwin "Buzz" Aldrin. Without spacecraft, none of these achievements would have happened.

What would Yuri Gagarin have done?

Yuri was the very first person in space. On April 12, 1961, he orbited Earth once in his ball-shaped spacecraft Vostok 1. Without this spacecraft, he would never have become world famous. But he could have carried on as a successful test pilot for the Russian Air Force.

What if there were no satellites?

We would have no satellite T.V. or satellite weather pictures, and mobile phones would not work very well. Ships, planes, and overland explorers could not use their satellite navigation gadgets. Without satellites, countries would have to find new ways to spy on each other. They could go back to the high-flying spyplanes used just after World War II, or use high-flying balloons carrying surveillance equipment.

WELCOME EARTHLINGS!

How much money would we save?

Space programs run throughout the world by different countries, like the Apollo Moon missions, have cost billions of dollars. Manned space flights are the most expensive type of missions. It has been estimated that NASA has spent over $80 billion on its manned space flights up to 1994, with nearly $45 billion spent on the space shuttle program alone! Even a single spacesuit worn outside the space shuttle costs $3.4 million!

Nonstick earthlings

Our everyday lives have been affected by the enormous technological leaps made during the age of space exploration. These "leaps" include nonstick coatings, used for lubrication in spacecraft. Also, the microtechnology needed in satellites has led to smaller and faster computers, some found in household appliances.

ELECTROMAGNETISM

The English physicist, Michael Faraday, discovered that electrical energy could be turned into mechanical energy (movement) by using magnetism. He used a cylindrical coil of wire, called a solenoid, to create a simple electric motor. He went on to discover that mechanical energy can be converted into electrical energy – the reverse of the principle of the electric motor. His work led to the development of the dynamo, or generator. You can make a powerful electromagnet by passing electricity through a coil of wire wrapped many times around a nail. Electromagnets are found in many everyday machines and gadgets. An MRI scanner (Magnetic Resonance Imaging), like the one pictured here, contains many ring-shaped electromagnets. With a solenoid and a current of electricity, you can close the cage.

CAGED!

1. Take a piece of polystyrene and edge it with cardboard. Stick plastic straws upright around three sides as the bars of the cage.

1

4

2. Cut out another piece of polystyrene of the same size for the roof of the cage. Attach a piece of plastic straw to the side above the door. Wind a piece of wire around a nail 50 times leaving two ends. Affix the nail to the roof, as shown.

2

3. Insert a needle into the straw so that it almost touches the nail. Cut out a rectangle of plastic for the door. Make a hole at the bottom of the door for the needle to fit through.

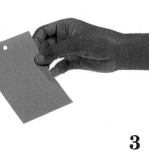

3

4. Stick a piece of cardboard across the door to help hold it open, and make sure the end of the needle just pokes through the hole. Now attach one of the wires to one terminal on the battery. Leave the other free. Make sure it will reach the other terminal. Put the animal into the cage.

WHY IT WORKS

When the current is switched on, the nail becomes magnetized as the current flows through the wire. The needle in the door of the cage is attracted to the electromagnet. As the needle is pulled toward the nail, the door closes to trap the tiger.

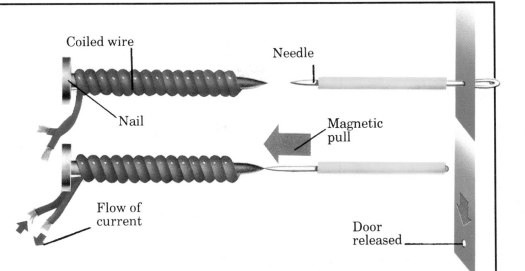

Coiled wire

Needle

Nail

Magnetic pull

Flow of current

Door released

BRIGHT IDEAS

Wind more turns of wire onto the electromagnet. The magnetic effect will increase. What happens if you use a more powerful battery? (Do not let it get too hot.)

Make another electromagnet using a shorter nail. This will also make the magnetic pull stronger.

Make an electromagnetic pickup by winding wire around a nail. What objects can you pick up? What happens when the current is turned off?

Use an electromagnet to make a carousel spin. Attach paper clips around the edge of a circular cardboard lid to be the roof. Make sure it is free to spin, and place an electromagnet close to the paper clips. The carousel should turn as you switch the current on and off quickly.

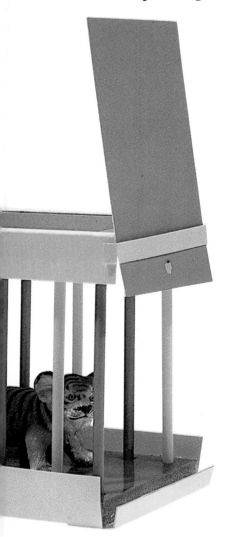

5. Now pick up the free wire. Allow the free wire to come into contact with the unconnected battery terminal. The needle should be pulled back toward the nail. The door will fall down, trapping the animal in its cage.

WHAT IF THE SUN WENT OUT?

Who turned off the lights? Why is it suddenly so cold? If the Sun no longer bathed our world in light and warmth, we might last a short time with fires, electric light, and oil or gas heat. But plants could not grow in the dark, and animals would perish from the cold. Soon all life would cease, and our planet would be dark and frozen. In fact this will happen, but not for billions of years. Our Sun is a fairly typical star, and stars do not last forever. They form, grow old, and either fade away or explode in a supernova, a massive explosion.

From the cradle to the grave

Throughout the universe there are massive clouds of gas, called *nebulae*. In some of these, the dust and particles clump together, and over millions of years, these clumps will form stars. Other nebulae are the wispy remains of a *supernova*, a star that has exploded.

What is a red giant?

An enormous human with red clothes? No, it is a star that has been growing and shining for billions of years, and is nearing the end of its life. As it ages, the star swells and its light turns red. Our Sun will do this in millions of years. It will expand to the size of a Red Giant, scorching our planet, before it explodes. Then all that will be left is a tiny white dwarf star that will slowly fade over millions of years.

How can we see a black hole?

When a really big star explodes, its core collapses and leaves behind a remnant whose gravity is so strong that nothing can escape its pull. This remnant is called a black hole. Because light cannot escape, it is impossible to actually see a black hole. However, its presence can be detected by the effect it has on objects around, such as gases, and waves, including light rays and X rays.

Great balls of fire

A typical star is made mainly of hydrogen gas. Huge forces squeeze together its atoms to form helium. This process is called *nuclear fusion*. As the atoms fuse they release energy, which radiates from the core, through the radiation zone. The energy is then carried to the surface by circular convection currents. Finally at the photosphere, the energy is radiated into space as light and other types of rays and waves.

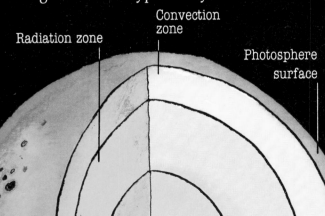

Radiation zone

Convection zone

Photosphere surface

Silent explosion

Sound waves can't pass through the vacuum of space, so we can't hear a star explode. But we can see it, as a glow that appears in the night sky, which then fades. It leaves behind a cloud, called a *nebula*.

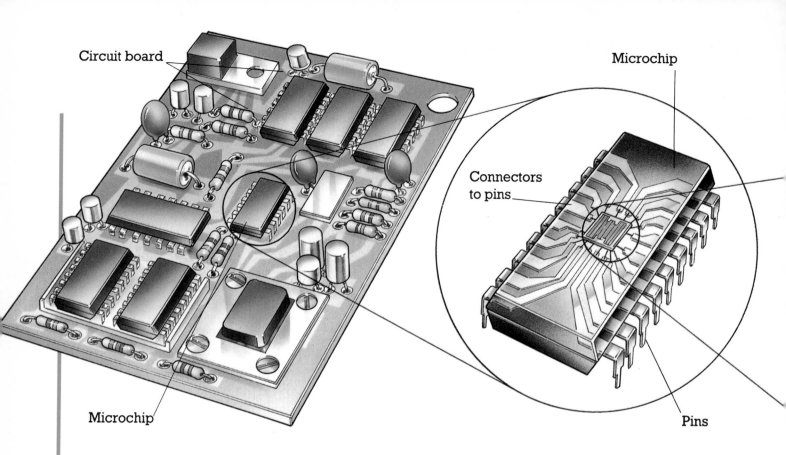

Circuit board

Microchip

Microchip

Connectors to pins

Pins

Inside a computer, electronic circuits are formed from components mounted on circuit boards (see above left), which are linked together by metal tracks on the boards. Many of the components are microchips (see above right). Each of them contains a tiny chip of silicon and this itself contains circuits composed of thousands of microscopically small components (see diagrams opposite).

THE MICROCHIP

Computers process information by first changing it into pulses of electric current that are then directed through complex electrical pathways or circuits. The majority of the electronic components on the computer's circuit boards are microchips. Most of them look like blocks of black or gray plastic with a row of metal pins along each side (see above right). The plastic block is to protect the chip which is buried inside, its metal pins connected to the metal tracks in the circuit board. The chip itself is often no bigger than a fingernail, although some are smaller. It is made from a slice of pure silicon on which intricately shaped layers of chemicals are added to form thousands of individual components. Silicon is one of a group of materials called semiconductors. Its resistance to an electrical current decreases as its temperature rises. This electrical resistance can also be changed by a process called "doping." This involves adding small amounts of different materials to the silicon. Some provide extra charged particles called electrons, forming n-type silicon. Others create a shortage of electrons forming p-type silicon.

Chips are made by adding specially shaped layers of different materials, such as aluminum, to a slice, or very thin wafer, of silicon. Each layer creates pathways for electric currents to flow through the chip. In the transistor illustrated on the bottom right of this page, a positive charge fed to the polysilicon gate attracts electrons from the p-type silicon base. This turns the transistor on as current only flows from the source to the drain when a gate current is applied. A negative charge at the gate repels electrons and turns the current and transistor off. Transistors commonly consist of three layers of silicon, either p-n-p or n-p-n.

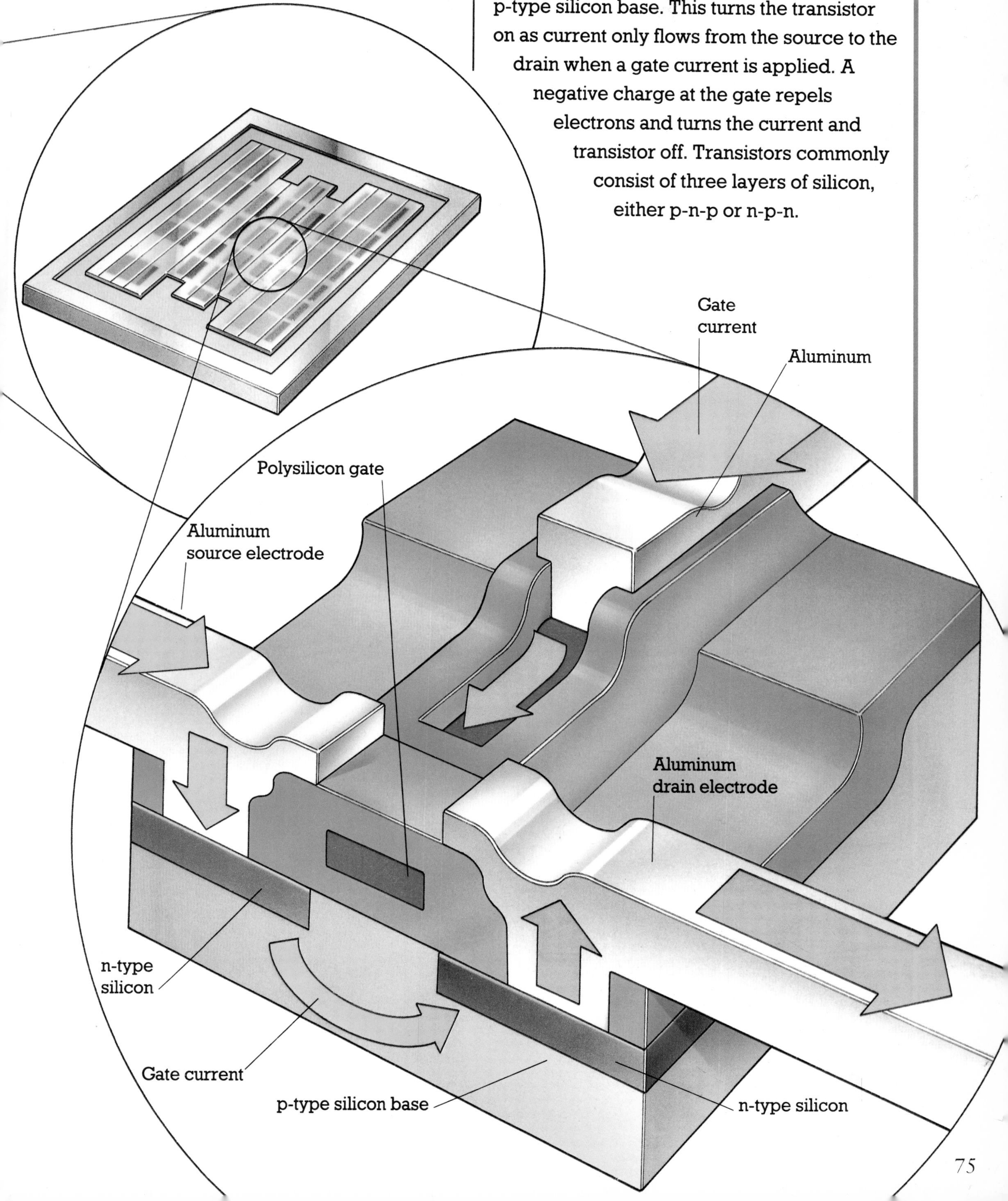

Gate
current

Aluminum

Polysilicon gate

Aluminum
source electrode

Aluminum
drain electrode

n-type
silicon

Gate current

p-type silicon base

n-type silicon

75

THE THINKING BRAIN

The main part of the brain that we use to think, decide and reason is the cortex – the thin gray layer on the wrinkled domes of the two cerebral hemispheres. The cortex looks the same all over. But brain research has "mapped" it to show its different parts are specialized for different jobs. We have maps on the brain!

PERSONALITY
Are you a good, kind person? Of course! The frontal lobes take part in the complex behaviors we call personality.

LEFT BRAIN, RIGHT BRAIN

In most people, the two halves of the cortex seem to have different tendencies. The right side is most involved in creative and artistic abilities such as painting, drawing, writing and playing music.

Artistic brilliance

Scientific excellence

The left side tends to take over in logical and rational thinking, as when solving mathematical sums, doing scientific experiments, playing chess and working out what to say.

SENSOR AND MOTOR

These two drawings show how we would look, if each part of our body was in proportion to the area of cortex dealing with it. One is for skin's touch, the sensory cortex. The other is for muscle movement, the motor cortex.

Motoring man   *Sensitive man*

MUSCLE CONTROL
The motor cortex is in overall control of the muscles, ordering them to work so that we can move.

THE INS AND OUTS
Information whizzes around the brain and body along nerves, as tiny electrical blips called nerve signals. Sensory signals come into the brain from the eyes, ears and other senses. Motor signals go out to the muscles.

TOUCH
The somato-sensory cortex is the "touch center." It receives information from all over our skin, about things we touch, and whether they feel hot or cold, or press hard, or cause pain.

SIGHT
The visual cortex receives and processes information from the eyes. It works out shapes, colors and movements, and identifies what we see. It is the site of the "mind's eye."

SMELL AND TASTE
The olfactory cortex sorts out smelly signals from the nose. The gustatory cortex is part of the touch area and receives tastes.

HEARING
Information from our ears, in the form of nerve signals, travels to the auditory cortex. Here it is sorted out and analyzed. We can identify most sounds by comparing them with sound patterns in our memory banks. For a strange or unusual sound, we may turn the head to see what has made it.

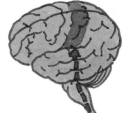

1 Signals come in from the senses.

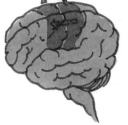

2 The brain decides what to do.

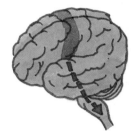

3 Signals go out to the muscles.

SPIES ON YOUR INSIDES

Another scanning method for looking inside the brain is PET (positron emission tomography). The PET scan shows where the brain is busiest and most active.

REFLECTION

When light rays hit a surface, they bounce off again, like a ball bouncing off a wall. This is called reflection. The way light behaves when it hits a reflective surface is used by people and animals to see more clearly. Cats have eyes designed to reflect as much light as possible, because they need to see in the dark. Inside high-quality periscopes on board submarines, prisms (blocks of glass) are used to bend beams of light around corners, making objects at the surface visible. Light can be made to reflect off a surface. Mirrors can also be used inside a periscope.
Make your own periscope and let nothing spoil your view!

UP PERISCOPE!

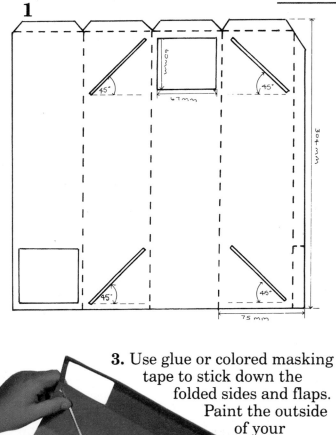

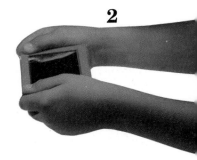

1. Use a ruler and pencil to measure and draw a plan of your periscope like the one shown here. Cut out the two windows and the four slits. Fold along the dotted lines.

2. Take two flat mirrors of the same size and put masking tape around the edges. These should be slightly wider than the periscope.

4. Slide the mirrors into the slits so that the reflecting faces are opposite each other. The edges of the mirrors will protrude from the periscope case. Make sure that they are secure. If they are not, they may slide out and break.

3. Use glue or colored masking tape to stick down the folded sides and flaps. Paint the outside of your periscope.

WHY IT WORKS

Light is reflected at the same angle as it hits an object, but in the opposite direction. The top mirror of the periscope is positioned to reflect light from the object downward to the other mirror. The bottom mirror is at the same angle as the top one and reflects the beam out of the periscope and into the eye.

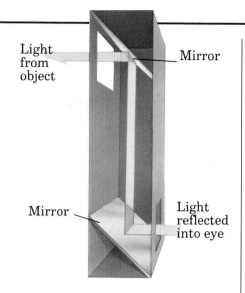

Light from object

Mirror

Mirror

Light reflected into eye

5. Use your periscope to view over an obstruction such as a fence or wall. Look into the bottom mirror to see what is hidden there. Notice what happens when you hold the periscope sideways. Try to look around corners as well.

BRIGHT IDEAS

☀ With three mirrors arranged in a triangular pattern, you can make a kaleidoscope. Cover one end with tracing paper and the other with cardboard. Make a hole in the oak tag to see through, and drop colored paper inside. Point the end toward a light source. Use materials of different colors.

☀ Write a message on paper and look at it in a mirror. You can turn a message into code by standing a mirror vertically above it and copying the image in the mirror. It can only be decoded with another mirror because it is upside down and back to front.

5

79

MOTORCYCLES

U.S. Militaire (1914)

The first motorcycle was a wooden bicycle fitted with a gas engine. Invented in 1885 by a German engineer called Gottlieb Daimler, it was only slightly faster than walking. By 1900 several firms were making motorcycles and, by 1914, the United States was taking the lead in technical development with bikes like the Militaire (*above*). Motorcycles were used by soldiers in wartime and are used by police officers and messengers today. But for many enthusiasts motorcycles are one of the most thrilling forms of transportation in the world.

Grooved tires increase grip

Twin-cylinder Royal Enfield bike (1960)

Honda Super Blackbird

TWINS AND "SUPER BIKES"

From the 1940s, some bikes were built with twin-cylinder engines for improved performance (*above*). Modern "super bikes" can perform as well as Formula-One racing cars. This Japanese Honda *Super Blackbird* (*right*) is the world's most powerful bike. Its top speed is 188 mph (300 kmh) and it can accelerate from 0-60 mph (0-96 kmh) in 2.5 seconds.

SCOOTER CRAZES

Scooters are an Italian invention, first produced by Vespa in 1947. These simple motorbikes are still

popular with young people in Europe. In the 1960s scooters were favored by teenagers called "mods." Some of them embellished their scooters with extra lights and mirrors (*left*). These "mod" bikes are now valuable collectors' items.

BIKE SAFETY
Safety is now an important issue in motorcycle manufacturing. Designers make handling easier by improving suspension and steering systems. Some of the most expensive bikes use air-sprung shock-absorbers to reduce the impact of bumps.

HARLEY-DAVIDSON

The famous American "Harleys" are ridden by the U.S. Army, most U.S. police forces, and thousands of other people – including the notorious "Hell's Angels" (*left*). The first Harley bike was built in 1907. In 1965 it introduced the Electra Glide – one of the most luxurious motorcycles ever produced. Recent models have a five-speed gearbox, disk brakes, and a rubber-mounted engine. The firm is one of the few U.S. motorcycle companies to survive competition from Japanese manufacturers today.

GOTTLIEB DAIMLER (1834-1900)
Daimler, a German engineer and inventor, is most famous for his automobiles. But he built the very first motorcycle before he built cars. His motorcycle was made mostly out of wood, and his son Paul rode 6 miles (9.5 km) on it to become the first motorcyclist.

MOTORCYCLE POLICE
Motorcycles are vital to the police (*left*). Motorcycle outriders escort ambulances and important people such as prime ministers and presidents.

SEATED SIDE BY SIDE

During World War II (1939-1945), Germany equipped motorcycles and sidecars with machine guns, while Britain and America used bikes mainly for staff work (*above*). After the war, sidecars were used before automobiles became widely affordable. Sidecars are now made in small numbers, and high-tech ones can cost as much as cars; this German model (*right*) costs $51,000!

FARMING QUADS
Quads are motorcycles with four large wheels used for riding across rough terrain (*right*). They are useful for farmers, but many people also enjoy the excitement of cross-country racing.

FEELING PAIN

When you stub your toe or bang your head, you might wish that the body lacked its sense of pain. But it is vital for survival. Pain warns you that the body is being harmed, and action is needed to get rid of the danger. Pain causes the body to make fast reflexes or movements called avoidance reactions. There are also pains that happen inside the body, indicating injury, pressure, or disease.

Sometimes, athletes, such as long-distance runners (above right), run into the "pain wall." This is when the body experiences so much pain that it feels that it cannot go on. When this happens, some athletes grit their teeth and continue running to take them through the "wall."

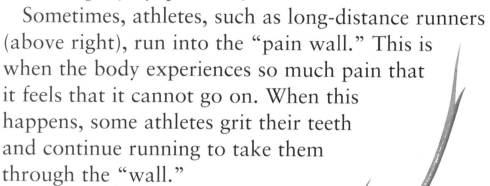

Nerve

Pain signal

Synapse

Dendrite

MUSCLE

NERVE CELLS
Neurons carry nerve signals (right). Each has a cell body, projections called dendrites, and a wire called the axon. The gap between nerves is called the synapse. Chemicals flood across this gap to the next neuron, to keep the signal going.

REFLEXES
A painful touch causes signals to go from the finger along the arm to the spinal cord. Here, they are sent to the brain, making you aware of pain. But already a reflex arc in the spinal cord has sent signals back to the arm muscles to jerk the finger away.

Axon

Cell body

BRAIN

Nerve signal going to muscle

SPINAL CORD

Pains can be described in terms of their characteristics – the way they affect you. The most common terms include stabbing, burning, crushing, and shooting (left).

Stabbing

Burning

Crushing

Shooting

There are many ways of reducing pain, including drugs such as analgesics which relieve pain, and anesthetics, which totally remove any feeling or sensation. Some affect the sensory cells and their nerves, called the peripheral nervous system. Others work on the central nervous system, which consists of the spinal cord and the brain.

NATURAL PAIN BLOCKERS
The body's own pain-control system uses substances called endorphins. These act in the brain and spinal cord to block pain signals at synapses between neurons (left). Endorphins are released when the body is active and stressed, such as when you are racing. This lets you concentrate on basic survival.

THE PHONOGRAPH

"And this week's number one in the pop charts, with over a million sold, is *Fifty-Three-and-a-Half*, by Dave 'n' Steve!" The record charts, along with vinyl records, cassettes and compact discs, and the whole area of recorded music and speech, began with "Mary had a little lamb."

It's him again – Thomas Edison. In 1877, he had the idea of recording sound so that it could be played back afterward.

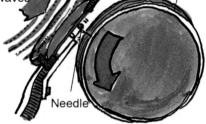

Sound waves   Diaphragm   Tinfoil

Needle

Edison designed a machine in which sound waves hit a thin, flat, metal sheet called a diaphragm. The waves made the diaphragm vibrate. The vibrations passed into a stylus or "needle." This pressed on a sheet of tinfoil, which was turning round on a cylinder. As the cylinder turned, the stylus pressed harder or softer for the different sounds, and made an up-and-down groove in the tinfoil. This was the recording.

Turning cylinder

Mouthpiece

Groovy sounds

The recording in the tinfoil was changed back into sound waves by reversing the process. The cylinder turned, and the grooves made the stylus vibrate. These vibrations passed to the diaphragm, which shook the air around it, to make the original sounds.

Horn amplifies sound

The machine was called a phonograph. Edison spoke into it for a test recording: "Mary had a little lamb." It worked!

Other inventors saw that the phonograph could be used for music and entertainment. In 1885, they used a cylinder covered with wax, instead of tinfoil. Lots of copies of one recording could be made, by pouring hot wax into a mold, which was shaped like the cylinder with its groove.

Cylinder to disc

In 1887, Emile Berliner came up with a flat disc, instead of a cylinder. The stylus was in a wavy groove and vibrated from side to side, rather than up and down as in Edison's version. Again, many discs could be made from one original recording.

Needle vibrates

Groove in disc

Berliner did not want his version to be confused with Edison's phonograph, so he called it the Gram-o-Phone. He gradually improved the sound quality. Soon people were buying the first recordings of songs. The "charts" had begun!

EDISON'S PHONOGRAPH

Drum

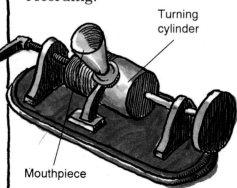

THE PHONOGRAPH

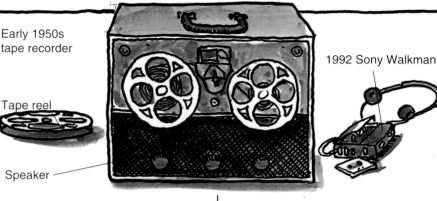

Early 1950s tape recorder

Tape reel

Speaker

1992 Sony Walkman

LPs to singles

Early record discs went round 78 times each minute, or 78 rpm (revolutions per minute). The discs were made of shellac. Vinyl records were introduced in 1946.

In 1948, the first successful long-playing records came out. They had much narrower grooves, went round at 33⅓ rpm, and lasted up to 30 minutes on each side.

Soon after, smaller vinyl discs came out. They usually had a single song on each side, they were seven inches across, and they went round at 45 rpm. They became known as "singles" or "45s."

BERLINER'S GRAM-O-PHONE

Horn

Turntable

Needle

Tapes big and small

In the 1950s, enthusiasts recorded sounds as patterns of tiny magnetic patches, on a long tape. The big tape reels were awkward to handle, and they took a long time to wind up.

In the 1960s, the Philips company brought out much smaller magnetic tapes, in little plastic cases or "case-ettes." They were neat and easy to handle. You could buy them already recorded, or make your own recordings. The cassette had arrived.

Tape cassette

In the 1980s, compact discs began to take over. They had patterns of microscopic bumps and pits, detected by a laser beam.

If Edison could see all the CDs, cassettes, LPs and hi-fi systems today, he would be amazed!

Pits

Handle

Laser beam

Aaaah, nice little dog

• The world-famous sign of HMV Records is a dog listening to the sound from a gramophone. This was a real dog, that lived in about 1900. Its owner had made a gramophone recording, but then died. When the recording played, the dog came over and sadly listened to – His Master's Voice.

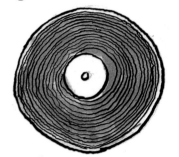

• How many grooves on a vinyl record? Two – a very long one on each side!

Compact disc

COLORS OF THE RAINBOW

Sunlight appears colorless but really it is made up of different colors. Sometimes you can see these colors — on the surfaces of bubbles or if there is oil on water. You may also see the colors across the sky in the form of a rainbow. In each case "white" light is being separated into different colors called the spectrum.

HOW A RAINBOW IS MADE

When the Sun comes out during a shower you may see a rainbow. The sunlight shines on the droplets of rain and gets separated into the colors of the spectrum. From a distance the light appears as a colored arc across the sky. People divide the rainbow into seven bands of color — red, orange, yellow, green, blue, indigo and violet. The colors always appear in the same order, with red on the outside and violet on the inside of the arc. The diagram shows how light which enters each raindrop is reflected, bent and separated into all the colors of the spectrum, which together form a rainbow in the sky.

△ It is impossible to reach the end of a rainbow — you can only see it shining in the sky at a distance.

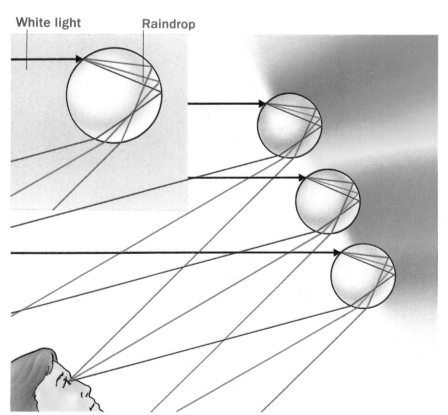

White light   Raindrop

MAKE A RAINBOW

You can see the colors of the spectrum by making your own rainbow. On a sunny day fill a pan of water and rest a mirror at an angle inside it. Stand the pan in front of a window so that sunlight falls onto the mirror. Then hold a piece of white cardboard in front of the mirror and move it around until you see a rainbow appear on it. You may have to move the mirror to get this right. The mirror and the water act as a "prism" — they separate white light into the colors of the spectrum.

THE NORTHERN LIGHTS

Sometimes dazzling displays of colored lights appear in the sky at night in parts of the world which are far from the equator. These lights are caused by huge explosions on the surface of the Sun known as "flares." During a flare, millions of tiny particles are sent out from the Sun. They travel very fast and some eventually reach the Earth's atmosphere. The Earth's magnetism bends the paths of the particles so they only reach the Earth's atmosphere near the poles. As they travel through the air they bump into other particles. These collisions produce light. In the North they can be seen best in parts of Canada, but they can also be seen in northern Scotland and Scandinavia. They are called the Northern Lights or "Aurora borealis." Similar lights can be seen in the South where they are called "Aurora australis."

△ The Northern Lights make an impressive display of color which looks like a constantly moving curtain in the sky.

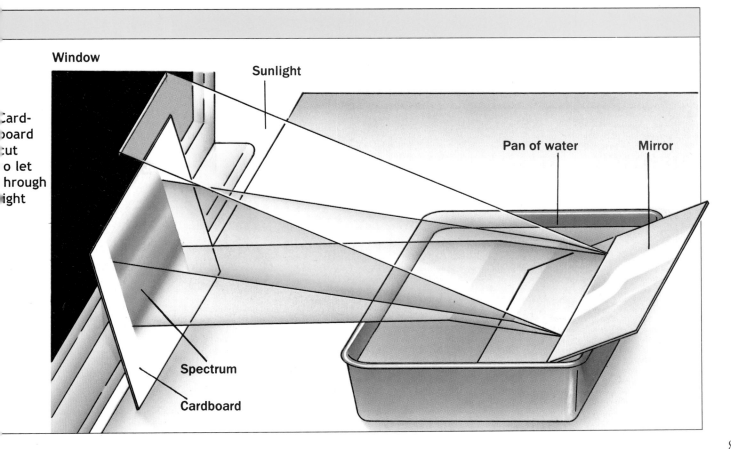

Window

Sunlight

Cardboard cut to let through light

Pan of water

Mirror

Spectrum

Cardboard

MAKING AN ASTRONAUT'S HELMET

This mask, a must for all space games, is based on a shell made of papier mâché. It is a vital piece of equipment when visiting alien worlds.

You could cut zigzag shapes from paper and glue them on to decorate the visor.

You could add a stars and stripes flag, painted on or made with pieces of straw.

Measure and pierce two small eyeholes in the mask with scissors. Alternatively, you could make the visor from transparent plastic so that you can see out more easily.

1 To make papier mâché, mix flour and a little water in a bowl, until you have a thick paste. Tear a newspaper into strips. Blow up a balloon and stand it in another bowl. Dip a strip in the paste, run it through your fingers to remove excess paste, and lay it on the balloon. Repeat this until the top half of the balloon is covered with at least three layers of newspaper. Leave it to dry overnight. Burst the balloon. **2** Trim the bottom of the papier mâché shape flat with scissors. Draw on the shape of the visor, and cut it out with scissors. **3** Cut two segments from an egg carton to make earpieces, and tape them on the sides of the helmet. **4** Cut out a circular piece of trash can liner to make the visor, and tape it inside the helmet. **5** The astronauts's eyes are two Ping Pong balls, and the nose is a cork. Cut eyebrows from an egg carton and tape all these features to the visor. Splay one end of a drinking straw, and tape it to the top of the helmet to make a radio receiver. Pierce a hole through the straw, cut another straw in half and push it through the hole, to form a crosspiece.

drinking straws

balloon

egg carton

Ping Pong balls

flour

STEP BY STEP

1

2

3

4

5

THE HARVEST

Traditionally, crops were harvested by hand *(left)*, either by picking them from trees and bushes or by cutting them with scythes *(right)*. This is still the case for some crops, such as tea, grapes, olives, and rice. However, massive combine harvesters usually do all the jobs needed to gather a variety of plants. They can do the work of hundreds of people in a fraction of the time and in most conditions.

Handle

ANCIENT HARVESTS

Images of the harvest are found among all ancient civilizations. They were often put in places of worship, as offerings to the gods in return for a good crop the next year. Many religions still hold harvest ceremonies.

Bronze Age sickle

THE SCYTHE

This consists of a long, straight handle made of wood with a metal blade, supported by a metal *grassnail*. Originally, scythes were sharpened with a *straik (below)* – a piece of pitted wood smeared with mutton fat and soft or stony sand, depending on the sharpness needed.

Grassnail

Iron Age sickle

Straik

CUTTING THE CORN

Scythes and sickles have been used since ancient times. The scythe dates from at least the Roman era and was used to cut a crop in a single, strong stroke. The sickle *(above)*, a smaller, lighter version, has been used since prehistoric times. Sickles are still used widely today, despite the development of machinery, because they are cheap to make and easy to use.

THRESHING

Threshing (separating) grains from stalks and winnowing (blowing) the husks from the grains were done by hand until 1780, when Andrew Meickle of Scotland built the first machine. Early threshers worked by horse power. The horses walked on a treadmill *(right)* to turn the machine. This was replaced in the mid-1800s by steam-powered engines.

FARMING... ON A MASSIVE SCALE!

China grows the most crops in the world – about 19% of the Earth's total production. This is followed by the United States with about 14%, then Russia. The most efficient harvest ever gathered was collected by an agricultural team in Britain. With a single combine harvester, they gathered 352 tons of wheat in only eight hours in August 1990!

1

PATRICK BELL

MULTI-PURPOSE MACHINES

In the early 1900s, work began on a combined reaper and thresher. It had an engine which worked the moving parts *(above)*. Today, combines come in many sizes, to suit different crops and fields. They have special attachments to harvest crops other than cereals, such as soybeans and cotton. Some are designed to harvest unusual crops like trees or potatoes. Most are self-propelled, although the smallest may be pulled by tractors.

HOW THE COMBINE WORKS
1. A cutting bar cuts the stalks with a moving knife.
2. The stalks fall onto a platform and are carried by a feeder to a threshing drum. This revolving cylinder separates the grain from the stalks, then the stalks are discarded and used for straw.
3. The grain passes through sieves and the husks are blown away by a fan.
4. The grain is fed into a tank. From here it is poured into trucks or sacks and taken away for storage.

4

2

3

MEDICINE

If you had a headache, would you let someone drill a hole in your skull to release the "evil spirits?" Medicine started like that! Most people nowadays expect doctors to cure them of life's aches and pains, and to heal serious diseases. Good medical care is sometimes not appreciated. It is just taken for granted.

Medicine began in the mists of prehistory. Some people found that they could cure an illness with a potion of plant juices, or a smear of animal fluids. They were the first doctors.

Some skulls over 10,000 years old have holes bored in them. The bone had grown back after being drilled, so these people must have survived after their "operation," trepanning.

A few early treatments worked, but many did not. Some were very harmful. Even so, if a medicine worked, people had respect and wonder for the doctor. As a result, doctors became powerful, and some were made into gods.

Trepanned skull

The father of medicine
One of the first proficient doctors was Hippocrates of Ancient Greece. He tried to rid medicine of magic and superstitions, and make it more scientific. He taught that a doctor's main aim was to help the patient, by finding the cause of an illness, and treating it. The results should be checked, so medicines could be improved. Hippocrates' main ideas are still followed today.

Hippocrates

Deadening pain
Surgery has been around for thousands of years. The only way of deadening the pain of the knife and saw was to get the patient very drunk on alcohol or to use opium. In 1842, Dr. Crawford Long operated on a patient using ether as an anaesthetic, to deaden pain and other sensations. Today we could not imagine even a very small operation, without an anaesthetic.

Joseph Lister

Killing germs on the body
Until the 1860s, patients who had operations often suffered and died, because their wounds became infected with germs. British surgeon Joseph Lister began to use antiseptics (germ-killing substances) to clean his operating instruments and the patient's cuts. Within a few years, surgery became much safer.

Alexander Fleming

Killing germs in the body
In 1928, British scientist Alexander Fleming discovered a substance which could kill bacteria (types of germs). It was made by a pinhead-sized mold called *Penicillium*, so he named it penicillin. This was the first antibiotic (bacteria-killing) drug. Many other antibiotics have been discovered, and they have saved millions of lives.

MEDICINE

Seeing inside the body

X rays were discovered by German professor Wilhelm Roentgen in 1895. People were amazed that they could pass through the body, except for bones. Soon X rays were showing up broken bones and suspicious lumps and bumps.

Doctors have many modern methods of seeing into the body. CAT scans and NMR scans show the inside parts in amazing detail. Thin tubes called endoscopes can be pushed into the body, to examine and photograph the insides.

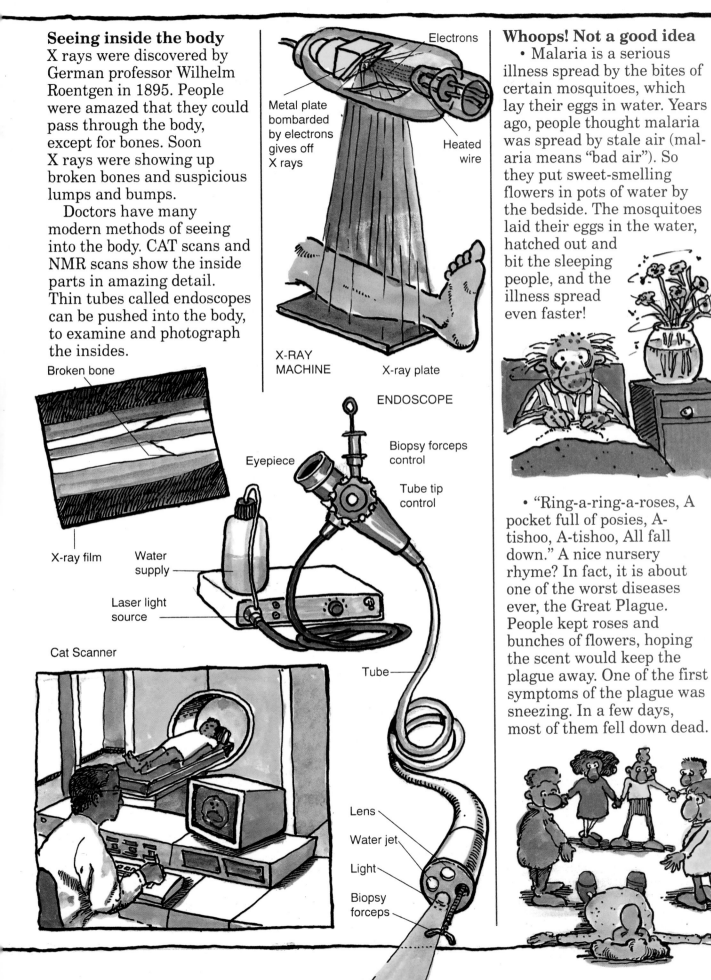

Electrons

Metal plate bombarded by electrons gives off X rays

Heated wire

X-RAY MACHINE

X-ray plate

Broken bone

X-ray film

Cat Scanner

ENDOSCOPE

Eyepiece

Biopsy forceps control

Tube tip control

Water supply

Laser light source

Tube

Lens

Water jet

Light

Biopsy forceps

Whoops! Not a good idea

• Malaria is a serious illness spread by the bites of certain mosquitoes, which lay their eggs in water. Years ago, people thought malaria was spread by stale air (malaria means "bad air"). So they put sweet-smelling flowers in pots of water by the bedside. The mosquitoes laid their eggs in the water, hatched out and bit the sleeping people, and the illness spread even faster!

• "Ring-a-ring-a-roses, A pocket full of posies, A-tishoo, A-tishoo, All fall down." A nice nursery rhyme? In fact, it is about one of the worst diseases ever, the Great Plague. People kept roses and bunches of flowers, hoping the scent would keep the plague away. One of the first symptoms of the plague was sneezing. In a few days, most of them fell down dead.

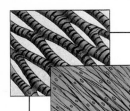

MUSCLES AND MOVEMENT

All the body's movements are powered by muscles. Muscle tissue is specialized to contract, or get shorter. The body has three main kinds of muscles. One is the skeletal muscles, attached to the bones of the skeleton, which you use to move about. There are more than 600 skeletal muscles, from the huge gluteus in the buttock to tiny finger and toe muscles. The other kinds of muscles are cardiac muscle in the heart (top left above) and smooth muscles in the stomach, intestines and other internal organs (left above).

Inside a muscle

A skeletal muscle has a bulging central part known as the body. This tapers at each end into a rope-like tendon, which anchors the muscle to a bone. As the muscle contracts, the tendons pull on their bones and move the body. The muscle body is divided into bundles of hair-fine fibers called myofibers. These long cells contain proteins that slide past one another to make the cell shorter in length.

Front shoulder muscle moves shoulder and upper arm.

Neck muscle moves head.

Biceps contracts and bends elbow.

Muscle sheath

Tendon

Myofiber (muscle cell)

Bundle of myofibers

Muscles in forearm bend fingers.

Front thigh muscle straightens knee.

Changing fashions

Bulging muscles have been in and out of fashion through the centuries. A few hundred years ago, plump bodies were seen as desirable. Today some men and women like to look slim. Other people work hard at body-building, training and lifting weights in the gym. They strive to increase the thickness of their muscle fibers through special exercises and diet.

Shin muscle bends ankle by pulling up foot.

Stories of the strong
Legends from many different cultures tell of well-muscled, strong men and women. Some are heroes, others are villains. Hercules of Ancient Greece had to undertake 12 "herculean" (very difficult) tasks or labors. In the Bible, the boy David fought and killed the giant Goliath with his slingshot. Samson was a hero who fought the Philistines, but he lost his strength when Delilah tricked him into having his hair shorn. Blinded and chained, he pushed the columns of the Gaza Temple and brought it crashing down on himself and his captors, as pictured right.

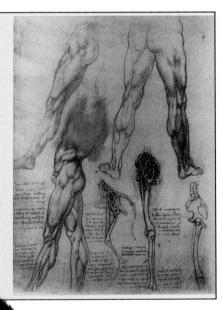

Triceps contracts, straightening elbow. Biceps relaxes and stretches.

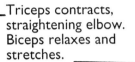

Master of art and science
During the Renaissance period, from about the 14th century, there was a rebirth of fascination in the beauty of the human form, and a scientific interest in the structure and workings of the body. Foremost in this field was the genius of art and science, Leonardo da Vinci (1452-1519). He performed amazing dissections of the body, especially the muscle system, and drew them with unparalleled skill and mastery, as shown here.

Biceps contracts, bending elbow. Triceps relaxes.

Quadriceps contracts, straightening knee. Hamstring muscles relax and stretch.

Quadriceps and hamstrings tensed to maintain crouched position.

Muscle pairs
A muscle contracts to pull on its bone. But it cannot do the reverse – actively get longer and push the bone the other way. So many of the body's muscles are arranged in opposing pairs, attached across the same joint. One partner of the pair pulls the bones one way, bending the joint. The other pulls the other way and straightens the joint, while its partner relaxes. Even a simple movement also involves many other muscles that keep the body balanced.

TOKAMAK
POWER FROM THE ATOM

Scientists in Britain and the United States are close to finding a way of controlling nuclear fusion reactions, like those that power the sun and the hydrogen bomb. If they succeed, nuclear fusion could provide an inexhaustible source of energy that would be far less polluting than either fossil fuels or conventional nuclear plants. The reaction they are trying to control takes place at high temperatures, between atoms of deuterium and tritium (both forms of hydrogen). Heated sufficiently, they combine to produce helium and a huge amount of energy. Just 10 grams of deuterium and 15 grams of tritium can, in theory, supply enough electricity to last the lifetime of an average person in a developed country.

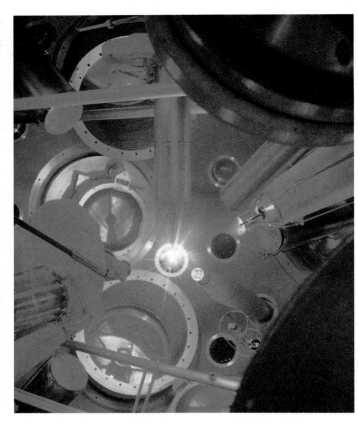

Lasers used to create fusion.

Another way of taming fusion is to make a tiny pellet of deuterium and tritium the target of a number of powerful lasers (left). When the lasers are fired together they are capable of reaching the very high temperatures needed to create fusion. The heating is so rapid that the atoms have no time to fly apart before they have fused together. American scientists have already had some success with this method, but there is still a long way to go.

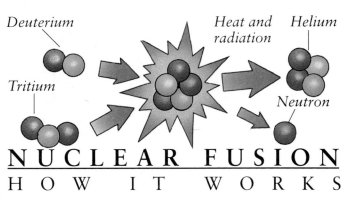

Deuterium

Tritium

Heat and radiation

Helium

Neutron

NUCLEAR FUSION
HOW IT WORKS

The diagram above shows how a fusion reactor causes the nuclei of deuterium and tritium to combine forming helium and a neutron, as well as heat. Nuclei usually repel one another, but if heated enough they will fuse together.

Scientists have created nuclear fusion in a tubular reactor that has been curved to form a continuous ring, known as a torus. The Russians were the first to try this shape, which they call a *Tokamak*. The atoms are injected into the ring and heated to about 180 million degrees Fahrenheit. The atoms try to escape, but a magnetic field holds them there long enough for them to start fusing together. So far, the most successful experiment, in the *Tokamak* at Princeton, has produced a short burst of energy. This was enough to prove to scientists that fusion will work. The European-funded Joint European Torus (JET), at Culham in Oxfordshire, England, has also produced brief, intense bursts of power. A practical power station will have to be much bigger than the *Tokamak*, and ways of removing the heat to generate electricity have to be found. The next stage will be an international project called ITER – International Thermonuclear Experimental Reactor – involving the United States, Europe, and Russia. If that succeeds, we could see the first fusion power stations in the first half of this century, perhaps by about 2020.

Scientists check the interior of the Tokamak Fusion Test Reactor *at Princeton, New Jersey (above). The fusion reaction takes place at about 180 million degrees Fahrenheit.*

Because no container could survive this temperature, a magnetic field keeps the atoms away from the walls. The by-products of the reaction are helium and high-energy protons.

The Princeton Tokamak *(right) holds the record for fusion output. In 1993 it produced a burst of power peaking at four megawatts and lasting seven seconds. So far the* Tokamak *experiments have consumed more energy than they created. There is still a long way to go before fusion becomes a viable energy source. Scientists have yet to find a way to release the energy slowly rather than explosively.*

The outside of the Princeton Tokamak.

WHAT IS LIGHT?

Light is a mixture of electrical and magnetic energy that travels faster than anything else in the universe. It takes less than one tenth of a second for light to travel from New York to London. Light is made up of tiny particles of energy called photons. The light moves along in very small waves that travel forward in straight lines, called rays. Light can travel through air and transparent substances, but can also travel through empty space. This is how sunlight reaches the earth. Light is similar to other forms of electromagnetic energy which have different wavelengths.

Radio waves
These have the longest wavelength. They are used for satellite communication and to carry TV and radio signals.

Visible light
Appears white or colorless, but is made up of colors, each with a different wavelength.

Radio	Micro	Infrared	Visible

Electromagnetic spectrum
This shows different kinds of electromagnetic energy arranged in order of their wavelengths. Wavelength is the distance between two consecutive waves.

Microwaves
Very short radio waves used in microwave ovens. They are also used in radar.

Infrared rays
Invisible rays, but we can feel the heat from them. They can be used to detect cancer and arthritis, or to take photographs in the dark.

Different ways of seeing
Most animals see visible light like we do. However, others have evolved sight which detects different wavelengths along the spectrum. Some insects see in ultraviolet light, which is invisible to most mammals and birds (except pigeons). Insect-pollinated flowers have lines which guide insects to their nectar. These lines only appear in ultraviolet light. The insects follow the guide lines to the nectar, scattering pollen on the way. This process helps the flowers to reproduce.

Bees cannot see the color red. They are strongly attracted by yellow and blue flowers which usually have strong ultraviolet markings. These markings attract bees to the flowers, and even pinpoint the location of the nectaries.

Ultraviolet rays

These cause us to tan and help the skin to produce vitamin D. Large amounts are dangerous and may cause skin cancer, although most ultraviolet light from the sun is absorbed by the ozone layer.

Gamma rays

These have the shortest wavelength. They are given off by naturally radioactive materials, such as uranium, and are part of the fallout after a nuclear explosion. They can travel through lead and cement, and damage living tissue.

UV	X rays	Gamma

X Rays

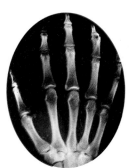

Called "X" by their discoverer because he was unsure of their nature. They pass through flesh, but are absorbed by bones and teeth, causing them to show up on x ray film. Small doses are safe, but large amounts are harmful to living tissue.

Taking the straight path

Try this experiment to prove that light travels in straight lines. Cut out two pieces of cardboard, about 8 in square. Make a hole in the center of each piece of cardboard, and stick a knitting needle through both holes, ensuring that they are aligned. Fix the cardboards onto a flat surface with modeling clay. Turn off the light and shine a flashlight through the holes. What happens? Try putting certain materials like cellophane, paper, or a book, between the two cards when the flashlight is on. Describe what happens to the ray of light.

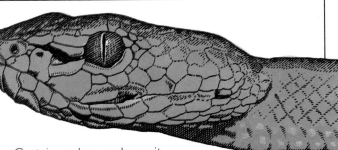

Certain snakes, such as pit vipers and pythons, detect prey by sensing the infrared light or "heat" they give out. They have special heat sensors which form an image from the infrared emissions. This helps them to hunt at night.

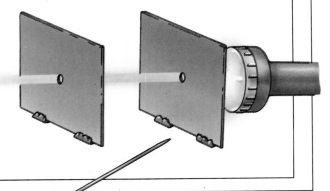

SPACE PLANES

TRAVELING AT MACH 18

Beyond the supersonic transport lies a new kind of plane altogether, one that combines flight and space technology to reach speeds of more than 10,000 mph.

The German space-plane design, Sanger, consists of a rocket piggy-backing on a hypersonic aircraft. The mother craft would carry the second stage rocket high into the atmosphere and up to a speed of Mach 16, where it would separate and go into orbit. The British version, called Hotol, has been developed by British Aerospace and Rolls-Royce.

These craft would have two jobs: getting satellites into space far more cheaply than shuttles or rockets can, and serving as passenger aircraft capable of getting to the other end of the earth in an hour or so. They would need new engines, and new heat-resistant materials for the fuselage. One idea used on the American National Aerospace Plane (NASP) is the scramjet, a jet engine that uses the speed of the plane through the air to ram air into the combustors, where it mixes with fuel and burns.

NASP would use liquid hydrogen as fuel for its scramjet engines. The hydrogen would be piped to the leading edges of the wings as a coolant. The main structure would be of titanium, but ceramics would be needed at the nose, where temperatures would reach 4,900 degrees F.

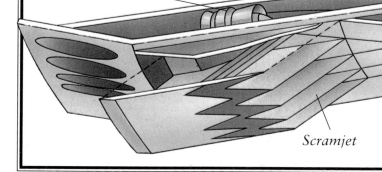

Turbojet

Scramjet

Space planes will need two types of engines. In this early NASA design (left), an ordinary turbo-jet engine is used for takeoff from the runway and acceleration to 2,000 mph. At this speed, scramjets can take over because the airflow is sufficiently fast for them to work. Scramjets could not be used alone, because they only begin operating effectively at high speeds.

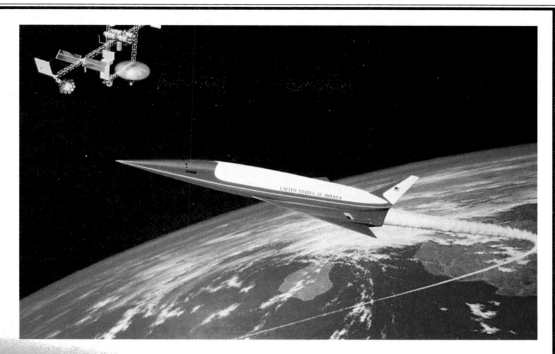

The NASP X-30 is intended to be a major element of the United States' planned permanently manned orbital space station, called Freedom.

Hotol's engine would burn liquid hydrogen and oxygen from the atmosphere to get it to high speed. Only as the air thins would it need to start burning the liquid oxygen carried on board. As a result, Hotol should be able to transport twice as much cargo as a shuttle.

Liquid hydrogen

Liquid oxygen

Payload

High-temperature insulation

Tank wall

Expansion

Most of the space inside Hotol would consist of a tank of liquid hydrogen. The tank walls would form part of the space plane's structure, linked directly to its outer skin (right). High-temperature insulation would be needed to keep the fuel tank cool. In wind tunnels, the Hotol design has been tested at speeds up to Mach 18 – the expected reentry speed.

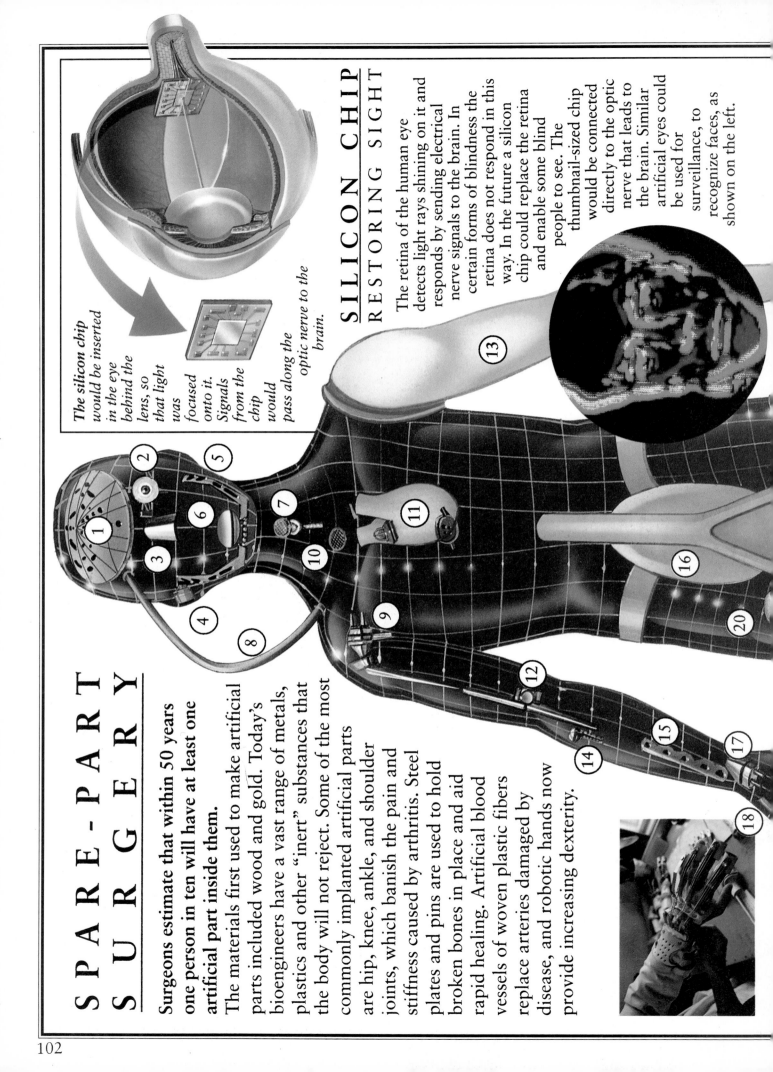

S P A R E - P A R T   S U R G E R Y

Surgeons estimate that within 50 years one person in ten will have at least one artificial part inside them.

The materials first used to make artificial parts included wood and gold. Today's bioengineers have a vast range of metals, plastics and other "inert" substances that the body will not reject. Some of the most commonly implanted artificial parts are hip, knee, ankle, and shoulder joints, which banish the pain and stiffness caused by arthritis. Steel plates and pins are used to hold broken bones in place and aid rapid healing. Artificial blood vessels of woven plastic fibers replace arteries damaged by disease, and robotic hands now provide increasing dexterity.

SILICON CHIP
RESTORING SIGHT

The retina of the human eye detects light rays shining on it and responds by sending electrical nerve signals to the brain. In certain forms of blindness the retina does not respond in this way. In the future a silicon chip could replace the retina and enable some blind people to see. The thumbnail-sized chip would be connected directly to the optic nerve that leads to the brain. Similar artificial eyes could be used for surveillance, to recognize faces, as shown on the left.

The silicon chip would be inserted in the eye behind the lens, so that light was focused onto it. Signals from the chip would pass along the optic nerve to the brain.

The chip contains hundreds of light-sensitive cells. All the cells operate at once, processing data very fast.

The latest artificial limbs are a huge improvement on previous versions. Here a champion at the Paralympic Games shows how the revolutionary "flex foot" enables him to run. The foot has a joint that bends (flexes) and then springs back. The latest artificial hands can be connected to nerves in the arm. Tiny electrical signals from the nerves control motors and levers, to reproduce some of the movements of real hands.

Above, a victim of the siege of Sarajevo is fitted with a new artificial hand.

Real hands can both grip and feel. The pincer movement between fingers and thumb is especially important for picking things up. Advances in the robot industry have produced artificial hands with touch-feedback that can grip items lightly or firmly. Some patients prefer simpler devices.

Artificial implants
1. Skull plate 2. Eye 3. Nose bridge
4. Hearing aid 5. Jaw plate
6. Chin implant 7. Electronic larynx
8. Valve to control water on the brain
9. Shoulder joint 10. Filter to prevent blood clotting in the lungs 11. Artificial heart
12. Elbow hinge 13. Artificial arm
14. Radial bone-bead 15. Metal forearm plate
16. Stoma appliance 17. Wristbones
18. Tendon 19. Thumb/wristbone connection
20. Hip joint 21. Femoral bone
22. Knee hinge 23. Artificial leg
24. Big toe

BOWS AND ARROWS

Bows were first used for hunting during the Stone Age, 250,000 years ago. The ancient Egyptians were the first people to use bows and arrows in warfare, in about 5000 B.C. Until the introduction of firearms in the 16th century, archers played a crucial role in combat. Bows were made from tough, flexible wood like yew. The equipment used in modern archery is derived from the shapes and qualities of medieval long bows.

EARLY ARCHERS

Stone Age people used arrows with flint arrowheads (above) to hunt large or fast-moving animals. The arrowheads, carved from pieces of solid rock, were deadly sharp.

ANCIENT WARFARE

The ancient Egyptians fought many battles against neighboring peoples, including the fierce Assyrians (left), in which both sides were armed with bows and arrows. The Assyrian Army, which was well-organized and its soldiers well-trained, eventually conquered the mighty Egyptian Empire.

MEDIEVAL ARCHERS

In the Middle Ages, bows were used by archers in the Norman army to help defeat the English at the Battle of Hastings in 1066 (right). During the late Middle

Ages (14th and 15th centuries A.D.), English archers used a more effective bow, the longbow (left), to win several battles against the French – including Agincourt in 1415. Arrows from longbows could pierce plate armor at over 110 yards, but using one took years of practice.

ROBIN HOOD

According to legend, Robin Hood was an outlaw who lived in Sherwood Forest, England, during the 12th century. He used his great skill with a bow and arrow to defend the poor against a corrupt sheriff.

ARROWHEADS

An arrow is made up of a wooden or steel shaft, with feathers or flights at one end (to make it fly straight) and an arrowhead (*right*) at the other. Hunting arrowheads often have barbs along the side, which catch in the animal. Bodkins are arrows designed to pierce steel plate.

General-purpose

Bodkin

Japanese

African (with barbs)

Indian

CROSSBOWS

The crossbow is a powerful and accurate bow, first used in the Middle Ages. Often the archer had to use a mechanical winding device to pull back the string (*above*). Crossbows are still used by troops today. This soldier (*below*) practices using one while wearing night-vision glasses.

COMPETITION BOWS
Today's bows (*left*) are made from artificial materials such as fiberglass, which is glued together in strips. They are very accurate and are used in sporting competitions, including the Olympic Games.

HUNTING BUFFALO
Native American tribes used wooden bows to hunt buffalo (*above*). They rode alongside the animals, so they could shoot them at short range. Buffalo provided them with food and with leather for quivers (*left*), which held arrows, as well as clothes and tents.

ESTIMATING TIME

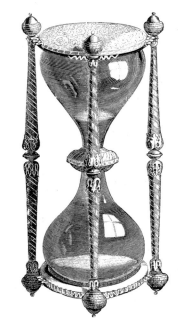

Hours are artificial units of time, first introduced by the Ancient Egyptians when they observed that shadows follow a similar pattern of movement each day. Shadow clocks and sundials were early time-measurement tools. In the 14th century, the sandglass, or hourglass, was popular, but it could only be used to estimate periods of time, varying from minutes to hours. It could not indicate the time of day. Onboard ship, a four-hour glass timed "watch" for the crew, until John Harrison invented the more accurate chronometer in 1735. This also calculated longitude and latitude. Early sandglasses were filled with powdered eggshell or marble dust.

TIME IS RUNNING OUT

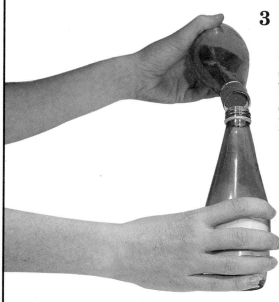

1 **1.** Wash and dry two small bottles thoroughly. Make an open-ended cylinder of cardboard and slide it over the top of one bottle. Cut out a disc of cardboard to fit inside and make a hole in its center.

2 **2.** Make sure the other bottle is absolutely dry. Now, carefully pour a measured amount of salt into it.

3 **3.** Position the empty bottle on top of the salt-filled bottle by sliding the cardboard cylinder over the neck of the lower bottle.

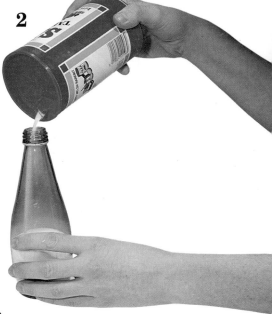

4 **4.** Check that the cardboard "seal" around the middle is secure. Carefully turn your timer over and observe what happens.

5. You can estimate the time taken for the salt to slide into the lower half of the timer by marking the side of the bottle with evenly spaced divisions. Use a stop watch to check exactly how long it takes for all the salt to slide to the bottom.

5

WHY IT WORKS

The upper vessel of the timer holds just enough salt to run through a hole, of a given size, in a given period of time. The force of gravity pulls the salt down through the hole and into the bottom container. The salt grains must be absolutely dry, so they don't stick. The size of the hole between the two vessels will determine the speed at which the grains will flow, but once established, the rate of flow will not vary. The total period of time depends on the quantity of salt.

Salt

BRIGHT IDEAS

Remove the regulator from between the two bottles and replace it with another, in which a hole of a different diameter has been made. Repeat this exercise a number of times, changing the size of the hole each time. What do you discover?

Design a sandglass that will run for exactly 3 minutes - use it as an eggtimer when you boil an egg.

Can you design another kind of sandglass that runs for a much longer period of time? Shape a funnel from cardboard, and insert it into the neck of a measuring container. Fill the funnel with sand and time how long it takes to pour through into the container below. Standardize your method of reading the scale - the top of the mound of sand will be concave.

Funnel

Sand

Scale

HELICOPTER ACTION

As any object falls through the air, the air pushes against it. Many trees have winged seeds that use this push to make them spin. The wings are shaped like airfoils, so as they spin they stir up low pressure above them. Higher pressure from the blanket of air below slows their fall as they drift away from the "parent tree." Helicopters also get lift from twirling airfoils. Their rotor blades spin so hard that the low pressure creates enough lift to carry them into the air.

WHIRLIGIG

1 Make two small airfoils from two pieces of oak tag, 4in by 3in. Fold each piece with an overlap.

1

2 Push the overlapping edges together and tape them together. One side will curve up like a wing.

2

3

3 Spread a little glue onto each end of a stiff straw or thin stick. With the curved side of the wings pointing up, glue them onto the straw so that they face in different directions.

4 Tape another straw to the one holding the wings. Use a piece of modeling clay to weigh down the end. This keeps the whirligig level.

4

5 To send the whirligig spinning, hold it between the palms of your hands. Brush your hands together, pulling one toward you and pushing the other away. As your hands come apart, the whirligig is released, twirling as it flies.

5

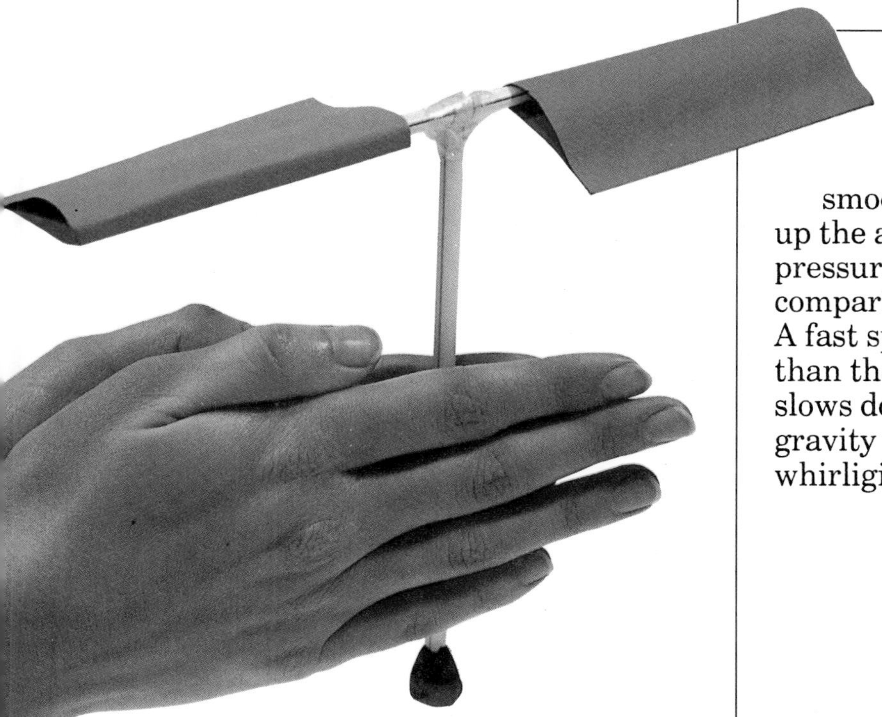

💡 Make another, bigger whirligig by doubling the size of the blades. Do larger blades give more lift because they create more low pressure? Therefore do you have to spin the whirligig as hard as before?

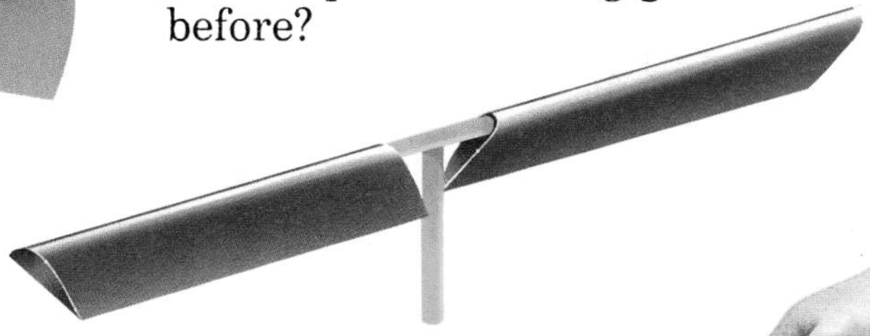

💡 Try fixing the wings of your whirligig at different angles. Notice how this affects the lift. Drop your whirligig. Does it spin as it falls?

WHY IT WORKS

As you twirl and release your whirligig the wings give it lift. Their airfoil shape cuts through the air smoothly, but the "bulge" in the top stirs up the air, creating low pressure above. The air pressure beneath the wings, higher by comparison, pushes the whirligig skyward. A fast spin creates a big lift, greater even than the pull of gravity. But as the spin slows down, the lift is lessened and gravity wins, pulling the whirligig to the ground.

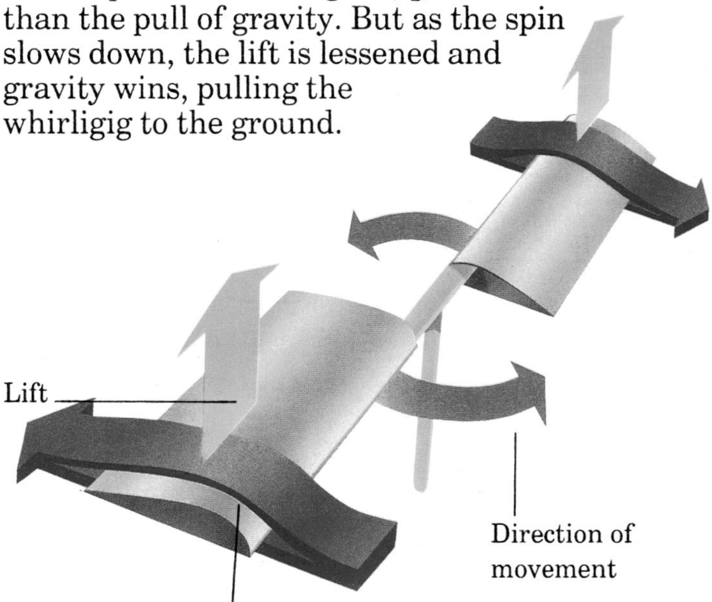

Lift

Direction of movement

Lower air pressure

SUNLIGHT

The Sun is a star that gives us light and heat energy. The Sun is about 93 million miles from the Earth. All plants grow toward the Sun. If you see a field of sunflowers, like the one pictured here, you will notice that they all face the same way, toward the Sun. Plants use the Sun's energy to make their own food. This energy is trapped by the green chlorophyll in a plant's leaves. During a process called photosynthesis, oxygen is released into the air as the sunlight is used to convert nutrients from the soil into food. The Greek word "photo" means light. Bioethanol is a fuel made by fermenting the food produced by plants like wheat. One day it could replace gasoline.

LEAVES

1. Half fill a shallow container with soil and scatter watercress seeds on the top. Keep the soil moist and place the tray in a sunny position. Leave it until the seeds sprout.

3. Leave the tray in its sunny position. You may have to wait as long as two weeks. Keep the soil moist while the cress is growing.

3

1

2

2. Cut out your initials from some cardboard, and place it over the seedlings. Make sure the sunlight cannot reach the plants beneath.

4

4. During this growing time do not remove the cardboard. You may want to turn the tray occasionally to allow an equal amount of light to reach every part of the tray.

5. When you observe that the watercress is fully grown, remove the cardboard. You should be able to see your initials in the seedlings. They will be a much darker green than the rest of the cress, where the light could not reach.

110

WHY IT WORKS

Sunlight is used by plants to convert nutrients from the soil into chemical energy for growth. When the leaves are covered, sunlight cannot be absorbed. No food can be manufactured inside the plant. Plants absorb carbon dioxide and water. These are converted by the green chlorophyll in the leaves into oxygen and simple sugars. The sugars are converted into food for the plant while the remaining oxygen and water is released into the air through small holes called stomata. These are located on the underside of the leaves. This process is called photosynthesis.

Sunlight

Carbon dioxide absorbed

Water absorbed

Oxygen and water released

5

BRIGHT IDEAS

Starch is produced when leaves photosynthesize. You can test for starch. Ask an adult to help you. Remove some cress from different parts of the tray and soak them in rubbing alcohol to remove any green chlorophyll.
Then place them on a clean surface and put drops of dilute iodine on the surface of each. Where starch is present, the leaves will turn blue, where there is no starch they will turn brown.
Plants always grow toward the sun. This is called phototropism. Plant a seedling in a pot and place it in a shoe box. Place a hole at one end of the box for the light to enter. The shoot will appear through the hole.

THE BODY SYSTEMS

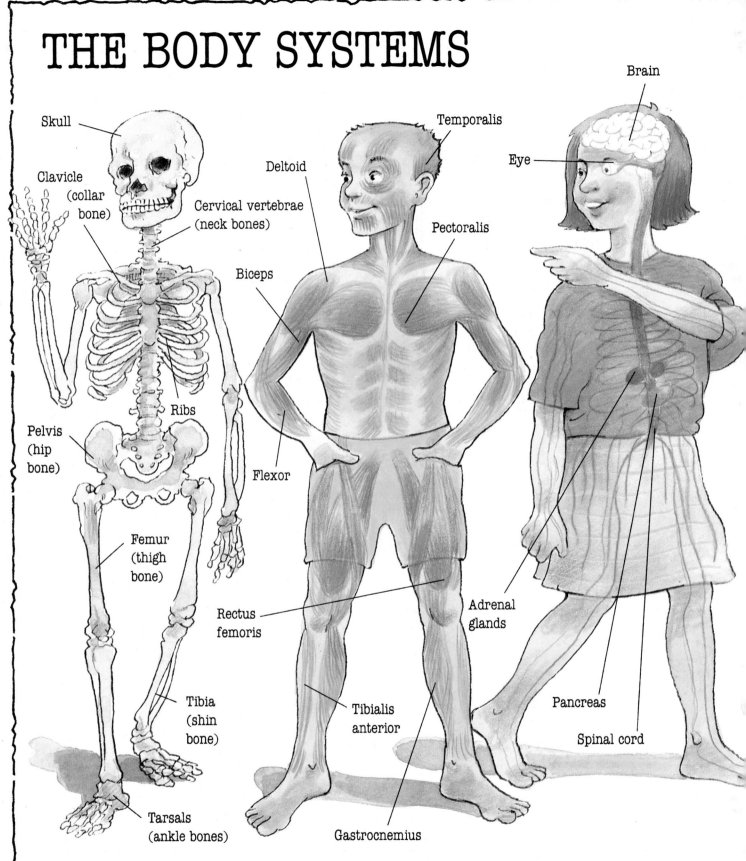

Brain

Temporalis

Skull

Eye

Deltoid

Clavicle
(collar
bone)

Cervical vertebrae
(neck bones)

Pectoralis

Biceps

Ribs

Pelvis
(hip
bone)

Flexor

Femur
(thigh
bone)

Adrenal
glands

Rectus
femoris

Tibialis
anterior

Pancreas

Tibia
(shin
bone)

Spinal cord

Tarsals
(ankle bones)

Gastrocnemius

The Skeleton
206 bones provide a rigid frame-
work moved by muscles and
protect soft parts like the brain.

Muscles
Over 640 muscles pull bones, so
you can move. Muscles are two-
fifths of our total body weight.

Nerves, Senses, and Glands
The nerves and glands control the
body's systems, using either
chemical or electrical messages.

Jugular vein

Aorta (main artery)

Heart

Windpipe (trachea)

Saliva gland

Liver

Gullet

Lungs

Kidneys

Ureter

Stomach

Large intestine

Femoral artery and vein

Bladder

Anus

Small intestine

Circulation

The body's cardiovascular system circulates blood through the blood vessels pumped by the heart. The blood spreads oxygen and nutrients, and collects any of the body's waste products.

Respiration and excretion

In respiration, the lungs absorb oxygen from the air, and excrete, or get rid of, carbon dioxide. The kidneys excrete wastes by filtering them from the blood, to form urine.

Digestion

The mouth, gullet, stomach, and intestines break down food and absorb nutrients into the body. The pancreas makes digestive juices, and the liver processes and stores nutrients.

SATELLITES
SPACE COMMUNICATION

**Communications satellites (comsats)
are capable of relaying computer
data, radio, telephone, and
TV signals across the world
in seconds.**

Today, there are hundreds
of comsats in orbit
around the Earth. Now
there are plans to build a
satellite network far
bigger than any of the
present systems. Made up
of 840 satellites, the network
would provide communications
services even to the remotest parts
of the world. The network would
provide a global information network,
linking computers in homes and offices.

*Consisting of 18
satellites 11,000 miles
above the Earth, GPS
works out the position
of the ship or plane by
calculating the time the
satellite's signal takes to
reach the
receiver on
earth.*

S A T E L L I T E S
HOW THEY ARE LAUNCHED

Comsats are launched into orbit by rockets
(above) or carried into space by space shuttles.
Satellites are equipped with booster engines that
guide them into orbit and help to keep them
there. Earth stations receive telephone, TV, or
radio signals and send them to the satellite. The
satellite amplifies the signals and retransmits them
to earth using a device called a transponder.

Equipped with portable satellite
transmitters, journalists can now use
comsats to send live broadcasts home
without having to rely on the facilities of
local TV stations. Aircraft and shipping
use a satellite navigation system called the
American Global Positioning System
(GPS). Portable receivers, such as the Sony
Pyxis, make GPS more widely available.

Equipped with portable satellite transmitters (right), newspaper and TV correspondents no longer have to relay live broadcasts via a local TV station with access to a satellite, neither do they have to rely on telephone links.

The Gulf War of 1991 provided an excellent opportunity to test the portable satellite transmitters in action (below). Reporting a war is difficult as well as dangerous, because normal communications are often disrupted. Portable transmitters could transform the coverage of wars and events in remote places.

SONY

PYXIS

GLOBAL POSITIONING SYSTEM

SONY

| POS | NAV | TRACK | EDIT | SET | MARK |

EXTENSION

CLEAR   RECALL

ENTER

GLOBAL POSITIONING SYSTEM   IPS-360

Astronauts maneuver a satellite into orbit from the space shuttle (above). Other spacecraft can launch satellites but only the space shuttle can be reused. Space shuttles have made it possible to retrieve satellites so they can be serviced in space. Those that need repairs can be brought back to Earth.

The Global Positioning *System was originally designed for military purposes, but it can also be used by sailors, climbers, explorers, and scientists. The small Sony receiver, called Pyxis (left) after the Pyxis or Compass constellation has opened up GPS to many more people. GPS allows users to know their position anywhere on earth within 30 ft, their speed to within inches per second, and the time to within fractions of a second. With the aid of GPS, scientists in Cumbria, England, have tracked the movements of individual sheep to see why some are still picking up radioactivity from the 1986 Chernobyl disaster.*

FOSSIL FUELS

Coal, oil, and natural gas are known as fossil fuels. They were formed from plants and the tiny creatures that lived on them, which were buried and subjected to the heat and pressure of the Earth over many millions of years. As energy from the Sun is stored in plant leaves as potential chemical energy (by photosynthesis), fossil fuels are a form of stored solar energy. The amounts available are large, but one day they will run out. Until then, we can use them as fuel and as raw materials for plastics, fertilizers, and chemicals.

Vincent Van Gogh

The painter Vincent Van Gogh (1853-1890) knew a lot about coal-mining communities, as he spent some time working as a missionary among the miners before becoming a painter in 1880. His early paintings were mainly of people in their work places, such as *The Return of the Miners* (below).

Coal deposits vary in location as well as quality. Coal can be extracted from near the surface by opencast mining. Often, however, the seam is deep underground and vertical shafts must be dug down to the seam. Tunnels or galleries are then cut into the seam as the coal is extracted.

SWOPS

When oil is found under the oceans, huge production platforms can be built to extract it. But if the well is small the cost may not be justified. Then a wellhead can be placed on the seabed, and oil pumped up to a tanker above, in a system called SWOPS – the Single Well Oil Production System. A single tanker, with room for 300,000 barrels, can serve up to three small fields.

seismic truck

the SWOPS system

The study of sound, or seismic, waves in the solid earth can help companies to locate oil and gas. Trucks can be fitted with equipment which sends controlled sounds into the earth.

Coal gas has now been replaced by cleaner natural gas, found in reservoirs in the Earth's crust, often with oil. It consists mainly of methane. The cheapest fuel available, gas is ideal for heating.

Gas lighting

In 1792 a Scottish engineer, William Murdoch, produced gas by heating coal in a closed container. By 1814 the first streets were lit by gas in London. However, really good lights were not available until 1885.

Swampland

Plants and animals flourished on Earth 300 million years ago. At that time, large areas of the Earth were swampy forests. These waterlogged conditions preserved the vegetation as part of its slow change to coal.

Crude oil is a heavy liquid that must be broken down, or refined, before being used. The process used to refine oil is distillation, and is based on the fact that different parts (fractions) of oil boil at different temperatures. The lightest fractions make gasoline, with heavier fractions being used for heavy fuel, oil, and tar.

Oil is less plentiful than coal, and is found where rocks prevent it escaping. Exploration rigs all over the world drill through ice and sea in the hope of striking oil. When they do, pipelines and tankers take it away. It is so valuable that its discovery can make a country very wealthy.

D.H. Lawrence

Coal mining was a major industry in the 19th century and produced mining communities around pits. It was a dangerous job which bred a strong community spirit. The novelist D.H. Lawrence (1885-1935), the son of a miner, wrote novels centered in these communities from his own life and experience.

WHAT IF THERE WERE NO PILOT?

Sometimes there isn't. At least, not a human pilot actually operating the controls. Many modern planes have an automatic pilot. It's not a robot sitting in the pilot's seat, but a set of controls incorporated into the main controls. The real pilot sets the plane's speed, height and direction, then switches to automatic, for a break. Of course, if something happens, alarms activate, and the real pilot takes the controls. In very modern planes, the computer-based auto-pilot can even take off and land the aircraft.

How do pilots "fly by wire?"

Computer screens are wired up to show speed, direction, engine conditions, and other information. Small levers and switches activate the flaps, rudder, and other control surfaces. This happens by sending electrical signals along wires to motors. This system is all controlled by the avionics system.

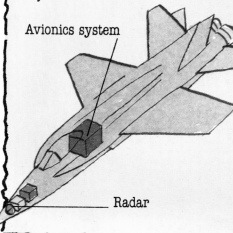

Avionics system

Radar

Tires, skis, skids, and floats

Airplanes can be equipped with a variety of landing gear, depending on their size and the conditions. Jet liners require wheels to withstand the pressure. Seaplanes need floats to keep them above water. Gliders and early rocket planes use skids, while planes that have to land on snow and ice use skis!

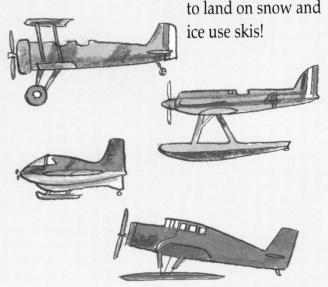

When can you see two sets of controls and instruments?

In the "head-up display." There are not really two sets. Part of the main display is reflected or projected upward onto the front windshield or canopy, or into the pilot's special helmet visor. The pilot can look ahead and see outside, and the controls at the same time.

What is a "black box?"

It's not usually black or box-shaped. It may be bright orange and cylindrical. But it's the usual name for an aircraft's flight data recorder. This device continually records the plane's speed, height, direction, and other information from the instruments, as well as radio signals and voice communications. It is specially made to be fireproof, shockproof, and waterproof. In the event of emergency or disaster, it can be recovered, and its recordings give valuable information about what happened.

Do you have to be strong to fly a plane?

Not really. Some controls are simple electrical switches and knobs. Others are levers, like the control column and rudder pedals, but they are well-balanced with counterweights and cables, so they aren't too heavy to move. But to fly a plane well, you do have to be alert and physically fit, with good coordination and quick reactions.

HOT ROCKS
POWER FROM THE EARTH

The earth's core is very hot – between 7,000° F and 8,000° F.

The heat is caused mostly by the breakdown of radioactive elements and, to a lesser extent, by heat left over from the earth's fiery beginnings. Tapping only a tiny fraction of this heat could produce large amounts of energy. In many places, where heat reaches the surface naturally in the form of hot water, geothermal power stations have been established. In the past decade, engineers have been trying to find ways to tap the earth's heat where it remains beneath the surface. This involves either drilling into underground sources of hot water, or creating them. To achieve this, holes are drilled into hot dry rocks, water is pumped down, then allowed to return to the surface hot enough to generate power.

People suffering from skin diseases bathe in hot water at the power station in Svartsengi, Iceland (below). Salt, clay, and algae in the water are said to be good for skin conditions.

Pipes (below) carry steam from the wellhead at Nesjavelljr in Iceland. The hot water produced heats the capital, Reykjavik. The steam is also used to heat greenhouses where fruit and vegetables are grown.

Hot-rock projects in the United States and Britain have produced some hot water, but less than was hoped. The British project, at a granite mine in Cornwall, lost a lot of water, and recovered less than one tenth of the heat expected. The American experiments in New Mexico found that the granite deep in the Earth could not be easily broken open.

In some places, hot water reaches the surface unaided. A geyser (right) erupts at Cerro Prieto geothermal plant in Mexico. Geothermal power plants create no pollution.

HOT ROCK PROJECT
H O W   I T   W O R K S

Hot rock projects (left), involve drilling two holes four to six miles down into rock that is up to 400°F hotter than the earth's surface. Granite is the best rock for this. A special drill bit is used to make the holes curve at the bottom like a "J." Water is pumped at high pressure into the deeper hole, causing the rock to crack. The water passes through the cracks to reach the shorter hole, and is heated as it travels through the hot rock. It emerges from the second hole heated to a temperature high enough to raise steam and generate electricity.

SUBMARINES

Most of the world's submarines are warships. They are designed to carry and fire torpedoes and missiles, or to lay mines, to destroy enemy vessels. Only the smallest submarines, called submersibles, are used for non-military, or civilian, work. Submersibles are built for repairing oil rigs, laying pipes on the seabed, and to study the underwater world.

There are two main types of submarine. The most common is the diesel-electric. This is powered by a diesel engine when on the surface of the water and by an electric motor when submerged. The second type is the nuclear submarine. This is powered at all times by a nuclear reactor. Diesel-electric submarines are the easiest and cheapest to build.

Military submarines are given code letters that describe their engines or the weapons they carry. For example, diesel-electric attack subs are coded SS, and nuclear powered ones SSN. Attack subs are designed to find and destroy enemy warships. Submarines that carry missiles which can be guided to a ship or target on land are coded SSG or, if nuclear powered, SSGN.

The nuclear powered hunter-killer submarine is fast and well-armed.

A submersible – for underwater research.

A nuclear missile-carrying submarine.

A diesel-electric submarine.

Submarines range in size from submersibles less than 6m (20ft) long to the 170m (558ft) Soviet Typhoon class submarine. This is nuclear powered and carries "ballistic" missiles (SSBN class).

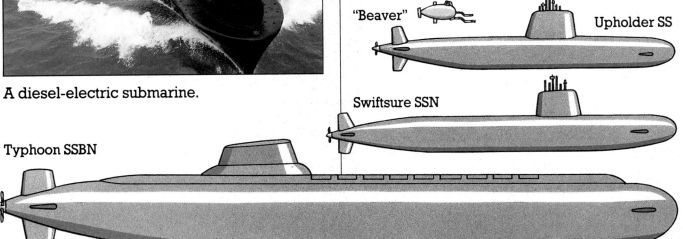

"Beaver"

Upholder SS

Swiftsure SSN

Typhoon SSBN

THE BITS OF THE BRAIN

The top half of your head is filled with a large lump of pinkish-gray, wrinkled looking substance, that feels like a mixture of pudding and jelly. But don't worry. All brains are like this. The human brain is the most amazing bio-computer. It can think, remember, predict, solve, create, invent, control, and coordinate.

BRAIN POWER

Are you smarter than a rabbit? Almost certainly. Your brain is much bigger than the rabbit's brain. Are you smarter than a sperm whale? Again, almost certainly, even though this huge beast's brain is five times bigger than your own. Intelligence is not just a matter of brain size. It depends on the relative sizes of the brain parts, and how they are connected. The cortex, the wrinkly gray part, is huge in the human brain. This is where intelligence, thinking, and complicated behavior are based.

SPIES ON YOUR INSIDES

The CAT (computerized axial tomography) scanner pictures a "slice" of the brain, with no discomfort or risk. It beams weak X rays through the head and displays the results on a computer TV screen.

Computer

CORTEX
The outer gray part, where thinking takes place.

X rays beamed from all angles as camera goes around head.

CEREBRAL HEMISPHERES
These are the bulging, wrinkled parts. They have gray cortex on the outside, and white nerves inside.

CORPUS CALLOSUM
This long bundle of nerves links the two halves of the brain, so the right hand knows what the left hand is doing.

THALAMUS
An egg-shaped area that helps to process and recognize information about touch, pain, temperature, and pressure on the skin.

LIMBIC SYSTEM
Sometimes called the "emotional brain," the wishbone-shaped limbic system is involved in anger, fear, pleasure, and sorrow.

HIPPOCAMPUS
Supposedly shaped like a seahorse, hence its name, the hippocampus is part of the memory system.

CEREBELLUM
This is like a mini-brain within the whole brain. It is vital for carrying out skilled, complicated movements, like doing a brain operation.

Stalk of cerebellum

PONS
This name means "bridge." The pons is a crossroads for nerves going up to the cortex, back to the cerebellum, and down to the spinal cord.

Medulla

Spinal cord

ELECTRICAL BRAINS
In 1800, Alessandro Volta of Italy invented the battery. He spent many years arguing with Luigi Galvani, who had discovered electricity while experimenting on animal nerves and brains.

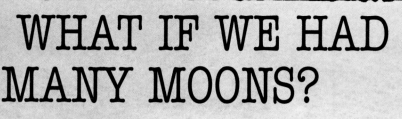

WHAT IF WE HAD MANY MOONS?

If the Earth had a lot of moons, night creatures might get confused. Moths use the Moon to find their way around – which would they choose if there were more than one? Owls, bats, and other night creatures might not wake up, as reflected light from the many Moons would keep the night sky bright. The Earth would also be more like the other planets. Most planets have lots of moons going around them. We have only one, which we call the Moon. At 2,160 miles (3,476 km) in diameter, it's much larger than most moons of other planets. With all these new moons we'd have to invent new names for them.

Is there a man in the moon?

No, but there were men on the Moon – the Apollo astronauts between 1969-1972. The patterns that we see on the Moon's surface, which resemble a crooked face, are made of giant mountains and massive craters. The craters, which can be as large as 625 miles (1,000 km) across, were made when asteroids and meteorites crashed into the Moon's surface.

The birth of a moon

Some scientists believe the Moon was probably formed at the same time as the Earth, from rocks whirling in space. Others think it was made when a planet crashed into the Earth, throwing up masses of debris, which clumped together to form the Moon. The moons of other planets may have been asteroids captured by the planet's gravity.

Which planet has the most moons?

At the moment Saturn has the most, with 18 moons as well as its colorful rings. This is followed by Jupiter with 16, and then Uranus which has 15. However, as telescopes get bigger and better, more moons may be discovered, so these numbers may change.

What happens if the Moon goes in front of the Sun?

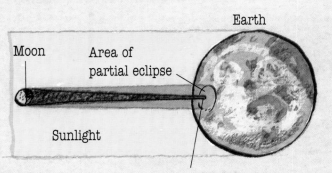

Moon | Area of partial eclipse | Earth

Sunlight

Area of total eclipse

It blocks out the Sun and casts a shadow on Earth, and we get a solar eclipse. But this does not happen all over the world. The total eclipse, with all the Sun hidden, is only in a small area. Around this is the area of partial eclipse, where the Sun appears to be only partly covered.

What's on the far side of the Moon?

The Moon goes around the Earth once every 27 days 8 hours. It also takes 27 days 8 hours to spin on its own axis. So the Moon always shows the same side to us. The far side of the Moon was first seen by the spacecraft Luna 3 in 1959, which sent back photographs of a lifeless moon, with no partying aliens!

SPEED AND ACCELERATION

Speed is how fast something is moving. Velocity describes both speed and direction. When a car turns a corner, speed may stay the same, but velocity changes. Speed describes how far an object travels in a period of time. For example, a snail moves at about 0.03 miles per hour, while Concorde travels about 1,300 miles per hour. A speedometer, like the one pictured here, indicates how fast a car is moving. Acceleration is how much the speed increases in a period of time. A decrease in speed is called deceleration.

AT FULL SPEED

1

2. Tape one end of the road to the narrow end of a shoe box. It can be lifted to different heights.

1. Cut the road from stiff cardboard as wide as a shoe box, but twice as long. Secure a small peice of cardboard in the middle of one side with a paper fastener.

2

3

3. Cut out one quarter of a circle. Divide it into angles of 10 degrees, and cut slits along the edge. Attach to the side of the box.

4. Check that the piece of cardboard on the road is in the correct position to slide into a notch. Pierce a hole in the lid of the plastic bottle. Cut off the bottom section. Tape over the hole, invert and fill with paint. Mount on top of the car.

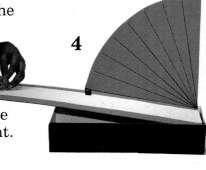

4

WHY IT WORKS

By observing the distance between the drops of paint on the inclined ramps you can estimate the speed of the vehicle. When they are close together, the speed is slowest. If the spaces are uniform, the vehicle must be traveling at a constant speed. If the spaces widen, the vehicle has accelerated, if they narrow it has decelerated. The bigger the mass of an object, the greater the force needed to make it move. When the slope is steeper, the truck accelerates faster. The spaces between the drops are wider apart toward the bottom of the slope. The speed can be calculated by dividing the distance by the time taken.

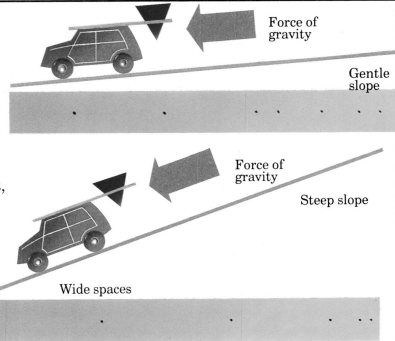

Force of gravity

Gentle slope

Force of gravity

Steep slope

Wide spaces

BRIGHT IDEAS

☼ Use a variety of toy cars on a sloping ramp and experiment with differing angles. Which car travels furthest? Did you use the same "push" each time to ensure a controlled experiment?

☼ Run the same car down a variety of angles, and allow it to run into a shoe box each time. Measure the distance that the box has moved. The distance that it moves depends on the speed.

☼ Try running on a beach at various speeds. Use a stop-watch to time yourself and measure the spaces between the footprints. Your footprints will be further apart, the faster you run.

5

5. Set the angle of the road. Place a long piece of paper over the road to record your results each time. Position the car at the top of the slope. Remove the tape just before it is released. Time each run accurately. Be careful not to push the car.

IN THE LUNGS

Deep in the lungs, something stirs. It is air, gently wafting in and out with each breath. The lungs' main air pipes, the bronchi, branch many times until they form hair-thin tubes, terminal bronchioles. These end in grapelike bunches of air bubbles, called alveoli. There are over 300 million alveoli in each lung. It is here that oxygen passes into the blood.

Terminal bronchiole

SLOW AND FAST
The brain constantly adjusts the breathing rate, according to the body's activity and oxygen needs. When we rest or sleep, the breathing rate is 15-20 breaths each minute. After running a race, it goes up to over 60 breaths each minute.

CHANGED AIR
Breathed-in air is about one-fifth oxygen, O_2. Coming back out, this proportion has changed to one-sixth. The difference is made up by carbon dioxide, CO_2, one of the body's waste products. Nitrogen, which makes up four-fifths of air, stays unchanged.

BLUE TO RED
Each alveolus air-bubble is surrounded by a network of microscopic blood vessels known as capillaries. Blue oxygen-poor blood flows into the capillaries. Here it picks up oxygen from the air inside the alveolus (as shown opposite), and turns into bright red, oxygen-rich blood.

Bronchus

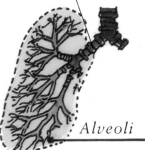

Alveoli

AIRWAY TREE
The bronchi branch 15 or 20 times to form the tiny bronchioles, with alveoli at their tips.

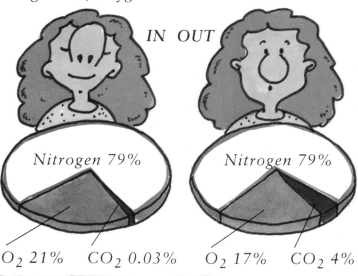

IN OUT

Nitrogen 79%   Nitrogen 79%

O_2 21% CO_2 0.03% O_2 17% CO_2 4%

BLOOD IN
Stale blue blood arrives at the alveoli along tiny arteries called arterioles.

BLOOD OUT
Refreshed red blood leaves the capillaries around the alveoli along miniature veins, venules.

DEEP-DIVING MOUSE
Deep-sea divers must breathe a special mixture of gases, including helium, due to the great water pressure. But breathing helium makes their voices sound like Mickey Mouse!

INSIDE AN ALVEOLUS
The wall of an alveolus is only one cell thick – which is extremely thin! The wall of the blood capillary is only one cell thin, too. So oxygen in the air inside the alveolus has hardly any distance to go, to get into the blood.

THE NON-STOP SWOP
Swopping the oxygen coming from the air to the blood, for the carbon dioxide going from the blood to the air, is gas exchange.

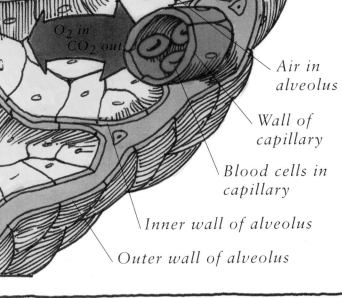

O_2 in
CO_2 out

Air in alveolus

Wall of capillary

Blood cells in capillary

Inner wall of alveolus

Outer wall of alveolus

LARGE AS OUR LUNGS
If a surgeon could iron all the alveoli flat, they would cover a huge surface, as large as a tennis court. That's a big area for gas exchange!

The First DIVERS

People have been diving in the sea for thousands of years, to search for valuable sponges and pearls or fish and other food. The earliest divers had no equipment, but with practice could dive to 60–100 feet (20–30 m) while holding their breath. As the centuries passed, curiosity and the hope of finding treasure or defeating enemies led to the development of all kinds of diving aids. Even the Italian artist and inventor Leonardo da Vinci (1452–1519) designed a device for breathing underwater, although he never tried it. Today, thousands of people enjoy diving as a sport.

Where do pearls come from?
A pearl is formed when an irritating particle of sand gets into a shell. It is covered in smooth *nacre* (mother-of-pearl) to stop the itch. Many shellfish can form pearls, but valuable pearls come only from tropical seas. The biggest pearl ever found weighs over 13 lb (6 kg) and came from a giant clam.

NATURE'S DIVERS
Human efforts to explore the oceans must seem puny to the great sperm whales. They can hold their breath for at least an hour and dive down to over 0.6 miles (1 km) to hunt giant squid.

JEWELS OF THE SEA
Some of the first human divers were pearl and sponge collectors. Pearls have been gathered in the Arabian Gulf since at least 3000 B.C. Early divers wore tortoise-shell nose clips to keep water out of their nostrils.

UNDER THE SEAS

The astronomer Edmund Halley invented the first diving bell in 1690. Divers sat in a wooden cask with an open bottom. As the cask was lowered, the air inside was squashed by the rising water, so extra air was pumped in from wooden barrels. Divers could walk outside the bell with small casks over their heads.

THE HELMET SUIT

The first diving suit was developed in 1837 by Augustus Siebe from Germany. The watertight rubber suit had a heavy copper helmet which kept the diver on the seabed. Air was pumped down from the surface. This allowed divers to work at over 300 feet (90 m) deep. A very similar but lighter suit is now used by commercial divers.

Getting the bends

Early divers suffered from a strange, often fatal disease – decompression sickness, or the bends. If divers surface too fast, the decrease in pressure makes the nitrogen gas in their blood form bubbles, which block the blood's flow. Dive computers can work out the safest ascent speed. Divers with the bends go into decompression chambers (right), with high air pressure to make the gas dissolve.

DIVING ALONE

In 1865, Benoît Rouquayrol and Auguste Denayrouze invented a diving set that did not need an air hose from the surface. Air was carried in a canister and fed through a valve in the helmet. But the set could be used only in shallow water at low pressure.

DIVING TODAY

Modern SCUBA (Self-Contained Underwater Breathing Apparatus) gives divers great freedom. The Aqua-Lung, the first breathing device to let people dive independently, was invented in the 1940s. The explorers Jacques Cousteau and Frédéric Dumas developed the demand valve, which gives air to divers when they breathe in (rather than all the time, which wastes air).

S W O P S
OIL PRODUCTION SYSTEM

When oil companies began extracting oil from beneath the world's oceans, they started with the biggest oil deposits, building huge platforms to drill from. But there are a large number of smaller deposits, containing under 100 million barrels of oil. Although the total amount of oil they contain is large, individually they are too small to justify a production platform. In the North Sea, the problem has been solved by introducing SWOPS – the Single Well Oil Production System. Two small oil fields, *Cyrus* and *Donan*, are now in production using the system which was developed by the oil company, British Petroleum.

Seillan, built in Belfast, is a 69,000-ton vessel that combines the jobs of oil tanker and oil production platform. Seillan can service three small fields.

134

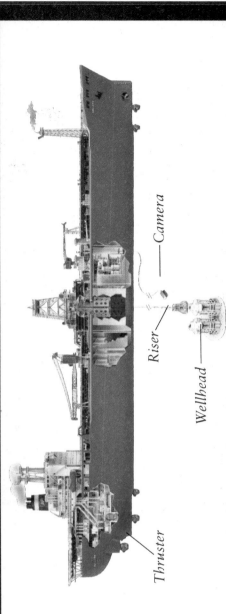

SWOPS involves drilling into the oil reservoir, and attaching a wellhead (which acts like a tap, or faucet) to the seabed. *Seillan*, a specially built production vessel, visits the wellhead, links up to it, and fills up with oil. When the oil emerges from the well, it is mixed with water and gas. *Seillan* is equipped with separators that can process the mixture into 15,000 barrels of oil a day. *Seillan* is kept exactly in position by seven stabilizing thrusters, powered by the separated gas. When its 300,000-barrel tanks are full, *Seillan* takes the oil to an onshore refinery. One ship can serve three small fields. Between visits, well pressure has a chance to build up again.

THE MOVING PLATFORM
HOW THE SWOPS WORK

Seillan positions itself over the wellhead and lowers a pipe, known as a riser, to the seabed. Seven thrusters on the ship keep it exactly in position. A T.V. camera guides the riser into position. The riser and the wellhead mechanically lock together, and *Seillan* fills up with oil, like a car at a gas station pump.

Camera

Riser

Wellhead

Thruster

YOU SMELL – IT'S NATURAL Each human body has its own distinctive odor. But exactly how much it smells is up to its owner. Too much of the wrong smell can make a person unpopular, because others avoid their nasty odor. This is where washing and hygiene triumph. Wash your body properly, regularly, and in all the right places, and you can avoid the dreaded B.O. Otherwise your best friend might not stick around for too long.

PROFESSOR'S FACT
WHAT IS B.O.?

• Body odor is a mixture of smells:

• Old sweat still on skin – not so much the sweat itself, but the bacteria that feed on it and then rot.

• Natural scents of sebum and skin oils, as they collect and become stale.

• Dirt, dust, grime, smears, and substances picked up by the skin.

 • All of the above rubbed into clothes that aren't washed. Sometimes it's not B.O. at all, it's C.O. – clothes odor.

INSENSIBLE SWEAT

Skin sweat glands make perspiration or sweat, to cool the body. The hotter you get, the more you sweat. But there's also "background sweating" that happens all the time, even in cold conditions. It's called insensible perspiration and it produces 1 pint (a half liter) daily. So even when you're cool, you're sweaty!

WHERE SWEATS MOST?

Sweat glands are more common in some parts of the skin, compared to others. So these areas sweat most, like the forehead, temples, armpits, palms, groin, backs of the knees, and soles of the feet.

Sweat cannot dry or evaporate from some parts, like the armpits, groin, and feet, since these are usually covered. Instead, it builds up in these areas – so they're the ones you should wash really well.

BIG TIP
NO COVER-UP

Perfumes, scents, and deodorants can help to cover up natural body smells, for a while. But they soon wear off, and they aren't the sole answer. There's no substitute for a good wash with water and soap. Put on the scent afterward, if you want.

THE WALKING ZOO You are covered in thousands of tiny animals, including fleas, lice, mites, ticks, worms, and other mini-beasts who are just trying to survive. They look for a meal of blood and fluids – and you might be the main course. These mini-monsters can get picked up from pets or farm animals, or a walk in the country or town. Most times, you know about them from an itchy bite or spot, where they bite to suck. Get rid of them quickly and easily, with a suitable soap or shampoo.

FLEAS

Most flea bites come from pet fleas. Not fleas that are pets, but fleas from pets, like cats and dogs. These long-jumping insects stray to human skin and suck a meal occasionally, but they usually soon move on. To solve the problem at the source, treat your pets and their bedding with anti-flea powders or sprays.

MICE LICE

Mice are tiny and pale. They live in hair and suck blood ... No, hang on, mice are small furry rodents. LICE are tiny and pale, live in hair, and suck blood. These insects lay eggs called nits, which are glued strongly to hairs. An anti-louse shampoo should get rid of them.

MICE MITES

These are arachnids, micro-cousins of spiders. There are hundreds of kinds. The scabies mite burrows into skin, lays its eggs, and causes REALLY ANNOYING ITCHING. These can be treated with a special soap to kill the poor little mites.

BIG TIP
FAMILY MATTERS

Sometimes people seem to catch skin pests again and again. It may be that other family members or close friends also have them, and the pests just get passed around. Ask the doctor or pharmacist, who'll advise the proper treatment. Then make sure other people in close contact get treated, too.

PISTOLS AND REVOLVERS

Pistols are much easier to handle than muskets and were used by officers, cavalrymen, highwaymen, and gentlemen in duels. Early pistols, called wheel locks, were replaced in the 17th century by flintlocks, which were cheaper and more reliable. During the 19th century, inventors began to design handguns that would fire several shots before the cylinder needed reloading. The first of these multi-shot weapons was the revolver, followed by the automatic pistol.

Cock

Trigger

FLINTLOCKS
To fire a flintlock pistol (*above*), the spring-mounted cock was drawn back. When the trigger was pulled, the cock was released and would hit the steel striking plate, producing sparks which ignited the gunpowder.

WILD WEST WEAPONS
The expansion of the American West in the 19th century coincided with the development of many new weapons, including breech-loading rifles and revolvers, such as the Colt .45, which were used by everyone, from cowboys (*above*) to soldiers, settlers, and Native Americans.

CRACK SHOTS

In the late 18th century, flintlock pistols began to replace swords as the main dueling weapon (*above*). Dueling pistols were always made in pairs and were of the highest quality to ensure accuracy. Today, the military and police practice accurate shooting at target ranges using modern pistols (*above right*).

HISTORICAL DUELS
Many important people took part in duels, including the Duke of Wellington, who defeated Napoléon at the Battle of Waterloo in 1815. The famous Russian writer Aleksandr Pushkin was killed in a duel in 1837.

Colt .45

REVOLVING CYLINDERS

During the 1830s and 1840s, several inventors began to develop handguns with cylinders containing five or six cartridges. The cylinder revolved so that the cartridges could be fired one at a time up the barrel. Only when the last cartridge was fired would the cylinder have to be reloaded. Colt (*left*) and Remington (*above*) were two of the most famous makes of the new handgun – called a revolver – which replaced the single-shot pistol and the slow, unreliable pepperbox pistol (*right*).

Remington

Pepperbox

SAMUEL COLT
(1814-1862)
Colt was an American inventor and gunmaker who carved an early version of a revolver out of wood. He produced the first working revolver in 1835, and introduced a system of mass production for the weapon.

THE LUGER PISTOL
Developed in the early 20th century, the distinctive-looking Luger pistol was adopted by the German Army and used in World War I and World War II (*left*). It could fire eight bullets in succession, from a magazine stored in the handgrip.

MODERN WEAPONS

The Israeli Desert Eagle (*above*) is a large, powerful, semi-automatic pistol, which uses a propellant gas to reload each time the trigger is pulled. Revolvers, too, are still widely used by armies and police forces: The highly powerful and exceptionally heavy .44in Magnum is perhaps the most famous since its use by the actor Clint Eastwood in the *Dirty Harry* movies (*right*).

Could we land on Venus?

Venus is similar in size to the Earth. But its atmosphere has clouds of corrosive sulfuric acid, and the surface temperature is 869°F (465°C). Not the place for a vacation!

Which planet is not named after a god?

All of the planets are named after Roman or Greek gods, except for Earth. It is named after the Old English word, "eorthe," meaning land or soil.

WHAT IF A SPACE PROBE TRIED TO LAND ON SATURN?

It would be very difficult, because there is hardly any "land" to land on! Saturn is the second largest planet, 75,335 miles (120,536 km) across, made up of mainly hydrogen and helium. A space probe would pass the planet's beautiful rings and disappear into the immense gas clouds of the atmosphere. As the probe fell deeper, the pressure would increase, and before long crush the probe. Farther down, the pressure is so great that the gases are squeezed into liquid. The planet's core is a small, rocky lump.

The planet of fire and ice

Mercury, the planet closest to the Sun, is only 3,048 miles (4,878 km) in diameter. Its atmosphere has been blasted away by powerful solar winds. This rocky ball has daytime temperatures ranging from over 806°F (430°C) – hot enough to melt lead – to a bone-chilling -292°F (-180°C)!

Stormy weather

Jupiter has a storm three times the size of Earth, about 25,000 miles (40,000 km) across. It's called the Great Red Spot, and drifts around the planet's lower half. A gigantic vortex sucks up corrosive phosphorus and sulfur, in a huge swirling spiral. At the top of this spiral, the chemicals spill out, forming the huge spot, before falling back into the planet's atmosphere.

Are there canals on Mars?

Not really. But there are channels or canyons. In 1877 Italian astronomer Giovanni Schiaparelli described lines crisscrossing the surface of the "Red Planet." He called them canali which means "channels."

What are planetary rings made of?

Saturn has the biggest and best rings – six main ones, made up of hundreds of ringlets. They are 175,000 miles (280,000 km) in diameter – twice the planet's width. They are made from blocks of rocks, ranging from a few inches to about 16 feet (5 m), swirling around the planet, and covered with glistening ice. Jupiter, Neptune, and Uranus also have fine rings.

Which planet is farthest from the Sun?

Pluto. No, Neptune. No – both! On average, Pluto is the outermost planet. This small, cold world is only 1,438 miles (2,300 km) across, with a temperature of -364°F (-220°C). Its orbit is squashed, so for some of the time, it's closer to the Sun than its neighbor Neptune. In fact, Pluto was within Neptune's orbit until 1999.

WHAT IS ELECTRICITY?

Electricity is an invisible form of energy which is stored in electrons and protons. These are the tiny particles in atoms (below) which make up all matter. Electricity is created when there is an imbalance of negatively charged electrons and positively charged protons. Current electricity is made up of moving electrons which travel through wires. In static electricity the electrons remain still. Electricity is a powerful and useful source of energy, but it can also be very dangerous.

Discovering electricity

Electricity was first discovered by the ancient Greeks, about 2,000 years ago. A Greek scientist called Thales noticed that a piece of amber (the hard fossilized sap from trees) attracted straw or feathers when he rubbed it with a cloth. The word "electricity" comes from the Greek word for amber – "elektron."

In 1600, William Gilbert (left), a doctor to Queen Elizabeth I of England, was the first person to use the word "electric." He carried out experiments and discovered that materials such as diamond, glass and wax behaved in a similar way to amber.

the atom's nucleus – made up of protons (green) and neutrons (red)

From the stars

The Sun and other stars send out radio waves through space. They are a form of electrical and magnetic energy which travel through space at the speed of light. They are picked up by huge dishes called radio telescopes. The radio waves are changed into electrical signals that give astronomers information about distant galaxies.

Investigating static electricity

Static electricity, used inside photocopiers and paint-spraying machines, can be generated by rubbing different materials together.

You can test materials for static charges, which are either positive or negative. Opposite charges attract things and like charges repel things. Experiment by rubbing different materials such as paper, plastic, metal, wood, and rubber with a cloth. Do they attract or repel things? Make a chart of your results.

Electrons everywhere

A particle accelerator (above) is used for research into atoms. By smashing atoms together, scientists have discovered over 200 particles, even smaller than atoms. A beam of electrons in an electron microscope (above left) enlarges objects millions of times.

negatively charged electron

Switch on the light

In fluorescent lights, an electric current makes gas glow. Neon gas makes red light, sodium gas yellow light, and mercury gas makes blue light.

Electricity for life

Most animals rely on electrical signals which provide them with information about their environment and control the way their body works. A network of nerve cells collects the information and sends out instructions. Invertebrates such as an octopus (right) have simple nerve nets.

Humans have more complex systems. The brain has an intricate network of nervous tissue (below). Our brain buzzes with tiny electrical signals, which trigger our heartbeats, to make our muscles move and sustain our body processes.

AIR RESISTANCE

If you open an umbrella and try to run with it on a calm day, you will find it difficult as the umbrella captures the air like a parachute, dragging you back. Whenever we move we have to push the air out of the way and we experience air resistance. Sometimes air resistance is helpful, for example in slowing down a parachute. It becomes a nuisance when it acts against a sports car. Some shapes are "streamlined" to move smoothly through the air. They experience less air resistance because the air does not rub against them too much and block their movement.

STREAMLINED SHAPES

1 You can test the force of air resistance. To make a fair test you need two of the same model cars. Make sure their wheels turn freely.

2 Cut two rectangles from a piece of cardboard. Again, to keep the test fair, make them the same size and shape.

3 Attach the rectangular cardboard to the front end of each car. Fold one smoothly over the top and bend the other one as shown. Tape them in place.

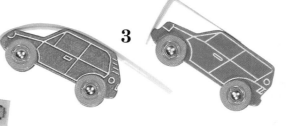

WHY IT WORKS

The shape of your cars makes them roll quickly or slowly. Air flows smoothly over the car with the rounded paper front. This streamlining allows it to roll faster than the car with the square front, which is held back by air resistance, or drag. Drag slows things down, creating ripples of air behind them. These moving ripples, or eddies, lower the air pressure behind the unstreamlined car, keeping it back as it moves.

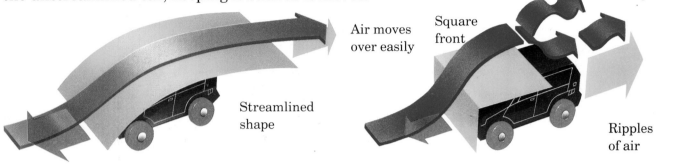

Air moves over easily

Streamlined shape

Square front

Drag

Ripples of air

4

4 Tilt a board on a book to form a ramp. Release the cars at the same time from the top of the ramp into a wind from a hair dryer. Notice which one experiences least air resistance.

BRIGHT IDEAS

Capture air with a simple parachute. Tie four strings to the corners of a large handkerchief. Fix a blob of modeling clay to the strings. Now make a larger parachute from paper, attaching the same piece of clay. It will drop more slowly than the first one because it captures more air.

VIRTUAL REALITY
COMPUTER ILLUSIONS

Computers have been used to create three-dimensional images for many years. "Virtual reality" allows us to take a walk inside those models, to make us feel that we are really there. Virtual reality technology uses computers to work directly on our senses – particularly vision, hearing, and touch – to create the illusion of reality, being in a computer-created spaceship or at Cluny Abbey. The user wears a special headset fitted with goggles. Computer-created images are sent to the headset. As the user moves, sensors feed data back to the computer, so that the view of the image changes, just as it would if you were moving through a real building or landscape.

In some virtual reality systems, it is possible to pick up imaginary objects, using a glove fitted with sensors that give the impression of gripping and lifting.

Virtual reality environments like the one above have many uses. Engineers have used virtual reality to plan telephone networks.

Architects have been using three-dimensional computer-aided design programs for several years. But a virtual reality design program like the one used to create Cluny Abbey would enable them to walk inside their buildings before they are constructed.

Virtual reality can be used for entertainment or for serious scientific research. Scientists at the University of North Carolina use it to build up molecules of drugs. Virtual reality not only allows them to see how atoms bind together but to feel when they don't.

The 1000SD is the first virtual reality computer game. Put on the headset and you find yourself in a computer-generated world. The design of the headset allows you complete freedom of movement. All your actions are controlled by a joystick.

The nave of Cluny Abbey, France (below), has been rebuilt inside a computer as an exercise in virtual reality. From the archaeological discoveries, and drawings made before it was knocked down, experts created the model with the help of IBM France.

CLUNY MONASTERY
RECONSTRUCTING THE PAST

Virtual reality is a powerful tool for archaeologists. It is now possible to re-create from plans and sketches what it felt like to walk through buildings long since lost. The monastery at Cluny, in southeast France, was a great center of culture and learning during the 11th and 12th centuries. What is known of Cluny comes from excavations during the 1900s. Virtual reality still has some way to go before it is truly convincing. The graphics alone present some big problems. The headset has to respond to your movement and to send images to your eyes at least as fast as a movie.

ENERGY CYCLE

Nearly all the energy we use comes originally from the Sun. It radiates through space and reaches Earth, causing plants to grow. These plants provide us with food energy in the form of crops and feed the animals which we eat as meat. They also provide fuel because plants and microscopic creatures that lived millions of years ago formed the fossil fuels – coal and oil. Rainwater, evaporated by the Sun's heat from the oceans, fills the rivers and provides hydro-electric power, while the wind is also produced by the Sun.

core

convection zone

radiation zone

photosphere

Sun worship
The Incas, Mayas and Aztecs, of South and Central America, worshipped the Sun with sacrificial offerings on Sun temples (left). The Incas thought the emperor was a descendant of the Sun. When he died his body was preserved and kept in his palace, where servants continued to wait on him.

an Egyptian Sun-god

In ancient Egypt, the Sun was one of many gods until king Akhenaten decreed that the Sun-god, Aten, should be the only god.

Myths and legends
In Greek mythology there were many gods. Phaeton, the son of the Sun-god Helios, was granted his wish that he should control the chariot of the Sun for a whole day. But Phaeton lost control of the horses and the Sun came too close to the Earth. He was about to burn the Earth up when Zeus, the chief god, struck him down with a thunderbolt.

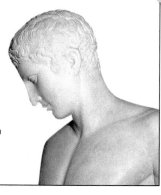

From the rim of the Sun, flares (above), called solar prominences, leap out. They are huge eruptions of hot gases linked to sunspots – slightly cooler areas on the Sun's surface.

The light of the Sun comes from the outer shell, or photosphere. Heated by nuclear reactions in the core, the photosphere glows brightly. The light from the Sun looks white, but it is a mixture of colors in the visible spectrum (above), or the range of colors visible to people.

Food energy

Different foods provide different amounts of energy, measured in kilocalories (or kcals). Each day we need 2,200 - 2,900 kcals from our food. A gram of fat contains nine calories; a gram of flour or sugar just under four; a gram of fish between one and two. Just sitting still uses 1.1 kcals a minute, while walking uses

3-4 kcals a minute and running fast about 15 kcals a minute. Sportspeople have a special diet to give them enough energy to compete.

Plants capture the Sun's energy by a chemical process – photosynthesis (left). A green chemical, chlorophyll, creates sugars from water in the plant's roots and carbon dioxide gas in the air. Other chemicals – nitrogen, calcium, phosphorus and potassium – come from the soil.

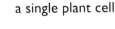

a single plant cell

Wheat worldwide

Six crops – wheat, rice, corn, barley, oats, and rye – provide about half of all human food energy. Wheat is used for making bread and is very effecient at converting sunlight into food energy, while for thousands of years rice has formed the basic diet of half the world's population. Wheat grows easily in cold, northern climates with low temperatures as opposed to rice, which needs warmer, wetter conditions in order to thrive and is grown on graduated terraces.

UNDER YOUR SKIN Like the rest of the body, skin is made of microscopic cells. On the skin's surface the cells are flat, hard, and tough for protection, like tiles stuck on a roof. As you move around, sit, walk, wash, get dry, and sleep in your bed, these cells are worn away and rubbed off your body. On average, you lose about 50,000 every second. Scratch your cheek – there go another few million! But don't panic, your skin won't disappear. Just under the surface, more cells are multiplying like mad, to replace the ones that have been rubbed off.

AT THE SURFACE

The outer layer of skin is called the epidermis. The hard, flattened cells on its surface are not very lively. In fact they are dead, ready to be rubbed away.

Hair

Sweat pore

Epidermis

Dermis

Blood vessels

Sweat gland

EVEN DEEPER UNDER THE SURFACE

The lower layer of skin is the dermis. It contains millions of microscopic touch receptors, nerves, blood vessels, hairs, and sweat pores.

BIG TIP
CARE IN THE SUN

Too much strong sunlight damages skin. In the short term, it causes the soreness and pain of sunburn. Over a longer period it may trigger skin growths and cancers. So Slip-Slap-Slop. Slip on a shirt or top, slap on a wide sunhat, and slop on the sunscreen lotion.

JUST UNDER THE SURFACE

At the base of the epidermis, cells are busy multiplying. They move upward, and get filled with keratin which makes them very hard. After three weeks they reach the surface to replace those that have been rubbed off.

PROFESSOR'S FACT
SKIN IS THIN

• Your skin's thickness varies on different parts of the body. On the soles of the feet it's more than 1/20th inch (five millimeters) thick. On the eyelids it's less than 1/100th inch (half a millimeter) thick and very delicate. So don't try walking on your eyelids!

KEEPING SKIN CLEAN'N'SHINY

HAVE YOU HAD A BATH TODAY? No? Eeek! Human skin is not self-cleaning, like an oven. It needs to be washed, all over. To do this, use soap and warm water, as found in most bathrooms. Splash the warm water onto your skin, rub on the soap to make a bubbly lather, scrub well, and wash off the sweat, dirt, and grime. If you've never seen soap before, it usually comes in small bars or lumps, often green or pink, sometimes with writing. Or use gel or a similar soapy substitute. If you don't use them, your skin will get dirty, grimy, sore, spotty, and smelly.

PROFESSOR'S FACT
HOW SOAP WORKS

• Tiny pieces of dirt clump together in larger sticky lumps, which you can see. Soap is a type of chemical substance called a detergent. It surrounds each tiny piece of dirt, making it come away from the main lump and away from your skin. Gradually the dirt clump is broken into millions of specks that float away when they are rinsed off.

TOP TO BOTTOM

Wash all over, not just the parts that show! Especially under arms, between legs, and in folds of skin. Sweat and dirt are more likely to collect here, trapping dirt and causing smells.

USING A SPONGE

Dunk the sponge in the water, and squeeze it to get the air out and water in. Then rub it on the soap, and rub it on you.

DON'T GET OLD SWEAT

Skin makes sweat, a watery, salty fluid with an important job – to keep the body cool in hot conditions. It also makes sebum, the natural waxy oil which makes skin supple and water-repellent. But as sweat and sebum dry, they become smelly and attract dirt. A good reason to bathe regularly.

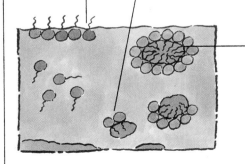

Soap specks float on water

Soap specks stick to a piece of dirt

Soap specks surround a piece of dirt

THE CAMERA

A camera is a device designed specially to record images on light-sensitive film. There are many different types of camera, but they all work in much the same way. The Single-Lens Reflex, or SLR, camera shown here is an example of one of the most popular types.

A camera is basically a lightproof box. A lens is fixed to one side and film is positioned inside the box opposite the lens. Light is prevented from entering the box by a shutter, a type of blind, behind the lens. When closed, the shutter stops light passing through a hole, the aperture, in the camera body. Light entering the lens of an SLR is reflected upwards by a mirror. At the top of this camera, a specially shaped block of glass called a penta-prism reflects the light out through the viewfinder. Other types of cameras have separate viewfinder and shutter-lens systems.

When a camera user wishes to take a photograph, or "shot", almost the exact image that will be recorded on the film can be seen in the viewfinder. At the right moment, the shutter is opened by pressing a button known as the shutter release. If the camera is an SLR, the mirror flips up out of the way, allowing the light to pass through the lens and reach the film. The lens bends the rays of light so that they produce a sharp image on the film. The amount the light rays have to be bent, or refracted, depends on how far away the objects are from the camera. Refraction is adjusted by rotating the focusing ring on the lens. Some cameras use a fixed-focus lens that is suitable for photo-

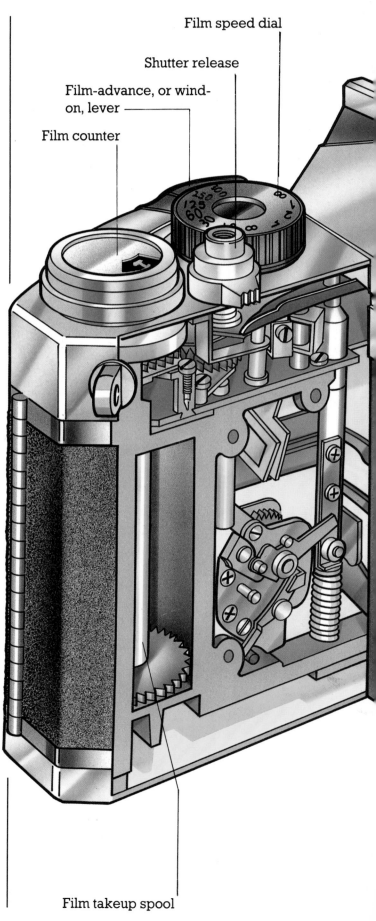

Film speed dial

Shutter release

Film-advance, or wind-on, lever

Film counter

Film takeup spool

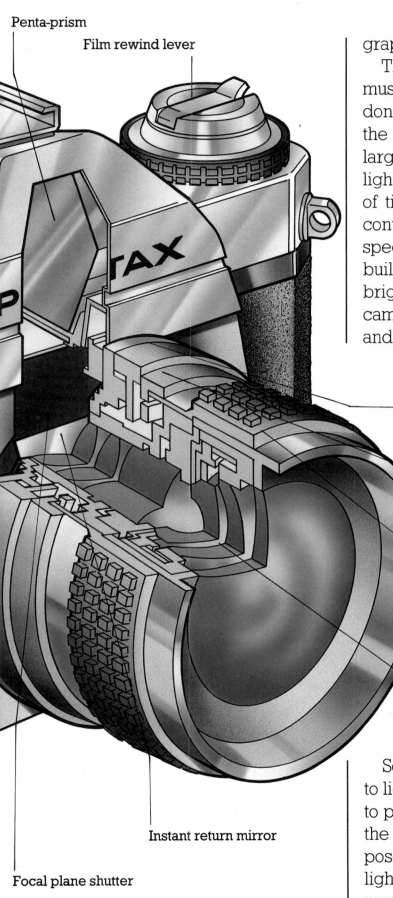

Penta-prism

Film rewind lever

P TAX

Aperture ring

Depth of field scale

Focusing ring

Lenses

Instant return mirror

Focal plane shutter

graphing both near and distant objects.

The amount of light falling on the film must be controlled carefully. This is done in two ways. The hole in the front of the lens, the aperture, can be made larger or smaller to vary the amount of light entering the camera. Or the length of time the shutter stays open can be controlled by changing the shutter speed. Most modern cameras have built-in light meters which measure the brightness of the scene in front of the camera. They use this to set the aperture and shutter speed automatically.

Some types of film are more sensitive to light than others. They need less light to produce a photograph. When setting the aperture and shutter speed to expose the film to the correct amount of light, the film sensitivity, or film speed, must be taken into account. High-speed film is the more sensitive.

CHEMICAL REACTIONS

Have you ever wondered why cakes rise in the oven or what the bubbles in soda are actually made of ? The answer in each case is the same – carbon dioxide. Carbon dioxide is a gas which is formed when two atoms of oxygen join with one atom of carbon. The formula for carbon dioxide is CO_2. There are actually small amounts of carbon dioxide in the air and green plants give out CO_2 during the hours of darkness. The CO_2 in cakes, however, is not drawn from the air. It is produced when acids react with carbonates or bicarbonates which are present in the ingredients.

VOLCANIC ERUPTION

1

1. Make a volcano which erupts with a foam of vinegar and baking powder. Form a cone shape around an old plastic cup or bowl. Glue the cone firmly to a base.

2

2. Cover the cone with 5 or 6 layers of newspaper and glue and leave to dry. Next, cover the base with glue and sprinkle with sawdust or sand. Paint the base and volcano.

3. When dry, a coat of sealing mixture (3 parts water to 1 part glue) will help to protect the paintwork when the volcano erupts. Once again, allow the whole thing to dry well before the next step.

4. Next, prepare the ingredients which will produce the reaction. You will need a small amount of baking powder or bicarbonate of soda and some vinegar mixed with a little red food-coloring.

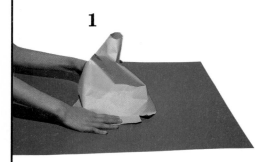

3

5. Put a teaspoon of the baking powder into the volcano then pour in a little vinegar. The reaction should be fast and a red liquid containing carbon dioxide will foam up over the sides of your volcano as it "erupts."

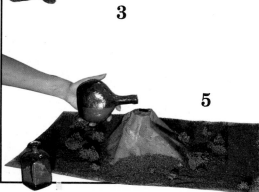

5

4

WHY IT WORKS

The vinegar is an acid and it reacts with the sodium bicarbonate (which is an alkali) in the baking powder. This reaction produces carbon dioxide gas. If you add water to baking powder you would also get a reaction, although it would be a slower one. This is because the baking powder contains acidic salts which become acids when the water is added. These acids react with the sodium bicarbonate to produce CO_2. It is partly this reaction which causes cakes to rise in the oven.

| Vinegar | Baking powder | Carbon dioxide |

BRIGHT IDEAS

Why not make your own fizzing lemonade? Mix the juice of 4 lemons with 1 quart of water then add sugar until your mixture tastes good. When you want a drink, pour out a glass, add half a teaspoon of bicarbonate of soda, stir, and drink at once!

6. Obviously, the more ingredients you use, the bigger the size of the eruption. A fizzing noise, known as effervescence, can be heard – this is the sound of the carbon dioxide gas being produced.

6

WATER AT WORK

From industry to agriculture to generating power for factories and homes, we make water work for us. To make the paper for this book, about 4 gallons of water were used, and other industrial processes, such as making cars, use vast amounts of water. Some power stations use water to generate electricity, while others need large quantities of water to cool machinery – you can often see water vapor escaping into the atmosphere through huge cooling towers (shown above).

Water power

Water generates power when it flows from a higher to a lower place. Waterwheels were originally used to capture the energy of flowing water and use it to turn millstones that ground corn or wheat into flour. Today, turbines use moving water to generate electricity. Modern turbines are huge machines weighing thousands of tons. They are usually placed at the bottom of a dam to make the best use of the energy made by falling water.

Waterwheel

The Itaipú Dam (shown above), on the Paraná River in Brazil, is one of the world's largest hydro-electric dams. Its 18 turbines can produce 13 million kilowatts of electricity.

Industry

In the United States, industry uses around 320 million gallons of water each day. Water is used for washing, cleaning, cooling, dissolving substances and even for transporting materials, such as logs for the timber industry (above). About 7,100 gallons of water are needed to make a car and eight quarts of water are used to produce just one quart of lemonade. The largest industrial users of water are paper, petroleum, chemicals, and the iron and steel-making industries.

Dam problems

Dams can cause problems for people and for the environment. Before a dam is built, people and animals have to be cleared from the area. If trees or plants are left to rot under the water, they make the water acidic and the acid may corrode (eat away) the machinery inside the dam. Reservoirs may become clogged by mud and silt which cannot be washed away downstream.

Irrigation

Crop plants, such as wheat or rice, need large quantities of water to grow properly. In places where there is not enough water, or the supply varies with the seasons, farmers irrigate the land. Most irrigation systems involve a network of canals and ditches to carry water to the crops. The sprinkler irrigation system (below) has an engine and wheels and moves across a field spraying crops with a fine mist of water.

Flood irrigation is used to grow rice. The fields of young rice plants are flooded, covering them in water. These fields are called paddy fields (above). It takes about 9,900 pounds of water to grow just one pound of rice.

An Archimedes screw lifts water up a spiral screw to a higher level. The device was invented by the Greek scientist Archimedes over 2,000 years ago. It is still used in some parts of the world today.

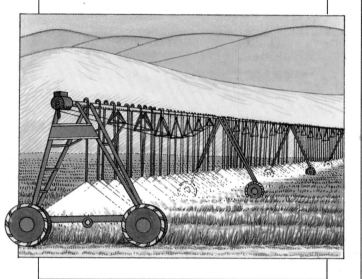

Hydroponics

Plants can be grown without soil using a watering technique called hydroponics. A carefully controlled mixture of nutrients are dissolved in water and passed over the roots of plants which are suspended in a tank of water. Hydroponics does not produce better or larger crops, but it is important in the study of plants and can be used in areas where soil is not easily available, such as on board a ship or in Arctic areas.

Solar salt

For centuries, salt has been a vital part in people's diets and has even been used instead of money. In countries such as China, India and France, salt is harvested from seawater and used for food flavoring or to make industrial chemicals. Seawater is left in shallow pools in the hot sun so the water evaporates, or disappears, into the air. Salt crystals are left behind and can be raked by hand or collected by machines. The salt is then taken to a refinery where it is crushed, ground and sorted before being packaged and sold. Evaporating seawater is the oldest method of obtaining salt. This kind of salt is called solar salt.

COMETS

From time to time an object looking like a star with a tail appears in the sky. This is a comet. The solid portion of a comet is a mixture of water, ice, frozen gases, and rock. Comets travel far out in space away from the planets in elongated orbits. When their orbits bring them close to the sun, frozen gases on the rocky body vaporize and form the bright tail which always points away from the sun. Some comets are well known. Halley's comet, for example, returns every 76 years. Amateur astronomers have been very successful in discovering new comets, so keep your eyes and notebooks open.

COMETS
A few comets are visible to the unaided eye. Others have to be viewed through binoculars or telescopes. Although many comets are well known, new comets are spotted from time to time and are always named after the person who first saw and recorded them.

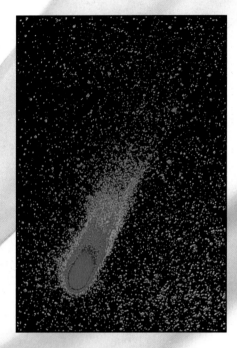

ORBITS OF COMETS

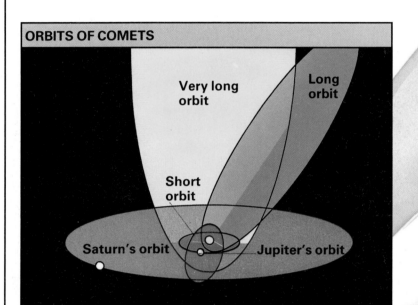

Very long orbit

Long orbit

Short orbit

Saturn's orbit

Jupiter's orbit

INSIDE A COMET
A comet's core is a mixture of water, ice, frozen gases, and rocks.

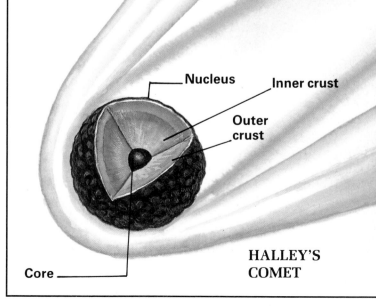

Nucleus

Inner crust

Outer crust

Core

HALLEY'S COMET

ASTEROIDS

There are thousands of pieces of rock and iron orbiting in our solar system. We call them asteroids. One group of vast boulders orbits the sun in a "belt" between Mars and Jupiter. Another group, the Trojans, occupies the same orbit as Jupiter. This includes the largest asteroid, Ceres, which is over 600 miles across. A third group orbits close to Earth. The Martian moons, Phobos and Deimos, may be asteroids that have been captured by the planet's gravitational field.

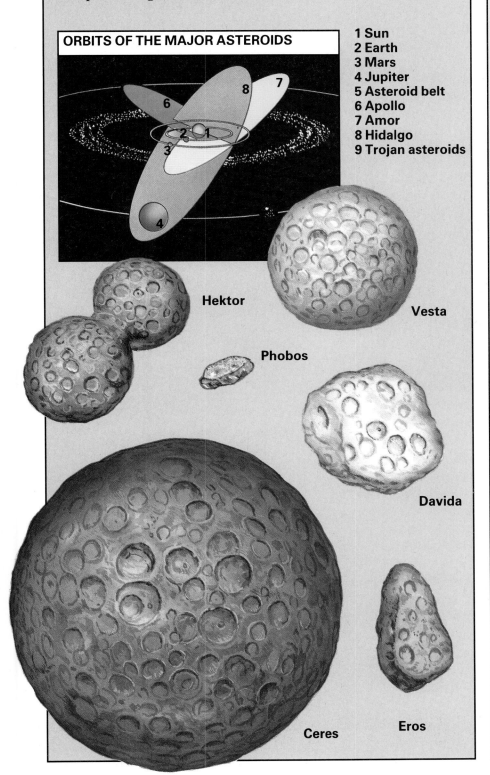

ORBITS OF THE MAJOR ASTEROIDS

1 Sun
2 Earth
3 Mars
4 Jupiter
5 Asteroid belt
6 Apollo
7 Amor
8 Hidalgo
9 Trojan asteroids

Hektor

Vesta

Phobos

Davida

Ceres

Eros

METEORS

Meteors, which are sometimes produced when asteroids collide, appear as streaks of light whenever chunks of debris from space enter Earth's atmosphere and burn up. Meteors can be seen on almost any night. Regular meteor showers also occur when Earth passes through a stream of particles left by a comet. The Orionid shower, for example, (October 16th-26th) is caused by particles from Halley's comet.

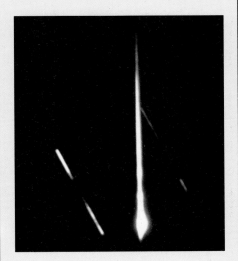

Because of their brief appearance, meteors are also known as "shooting stars."

ROCKS FROM SPACE
A meteor that reaches Earth without burning up is called a meteorite.

THE PRINTING PRESS

If printing had never been invented, then

...Exactly!!

The year: 1455. The place: Mainz, Germany. The person: Johann Gutenberg. The result: about 300 copies of what we now call the Gutenberg Bible, the first full-length book made on a proper mechanical

Johann Gutenberg

printing press. It had 1,284 pages, two columns on each page, 42 lines in each column. Fewer than 50 copies survive.

We see printed words and drawings and photographs not only in books. They are in newspapers and magazines, on containers and labels, on billboards, even on T-shirts, almost everywhere! Printing is central to the way we learn. Imagine life without school textbooks! (Then again, don't.)

One copy at a time

There was printing five centuries before Gutenberg. In China, shapes of pictures and writing were cut onto wood or stone blocks. Printers spread ink on the block, pressed paper on it, and the ink on the raised bits stuck to the paper. They could make many prints from one block.

Early printing block

But each new work needed a new block. So a system was developed with small metal blocks, which could be moved and arranged in different orders.

Early Chinese

However, the complicated Chinese writing system, with thousands of symbols, meant slow progress.

There were also hand-made books before Gutenberg. Monks, especially, spent years writing and decorating the letters by hand, with pens, one beautiful book at a time.

Moveable type

Gutenberg and his helpers produced hundreds of small metal blocks, each with a raised part of one letter or symbol, in reverse. The letters could be chosen and arranged in a frame, to print copies of a page from the book. Then a new set of letters was put into the frame for the next page. And so on, one page at a time.

Metal type blocks

GUTENBERG'S PRESS

Inking the type

THE PRINTING PRESS

Gutenberg Ink Inc

Another Gutenberg advance was new ink that stuck to the metal type, rather than to the usual carved woodblocks. And another was a powerful pressing machine that squeezed the paper against the inked type. It was adapted from a winepress, for squashing grapes!

Because the raised, inked letters were pressed onto paper, the method was called letterpress.

The letters and symbols were called type, and setting them up correctly was typesetting. For a long time they were placed by hand. In 1884 the Linotype machine made a line of type from molten metal that quickly went solid.

The press

Finished copies

Typesetting

Printed page

MODERN PRINTING PRESS

Computer

Paper feed

Yellow ink feed

Printing plate

Offset roller

Cyan ink feed

Magenta ink feed

Black ink feed

Friedrich Konig invented the rotary or cylinder press in 1811, saving time winding the press up and down for each paper sheet.

Gravure and litho

In gravure or intaglio printing, the areas to print are not higher than their surroundings, but lower. The rest of the ink is scraped or wiped away before the paper is pressed.

In lithography, the printing surface is flat. But the areas to be printed have their surface changed, so that the oil-based ink sticks only to them, and not to the untreated areas around. "Litho" was invented about 1796 by Aloys Senefelder. It is now the most popular kind of printing – this book was printed using litho.

Color pictures are printed from tiny dots of four ink colors: yellow, cyan (greeny-blue), magenta (reddy-purple) and black. Look at this one under a magnifying lens.

The Age of Learning

Within 50 years of Gutenberg, there were printers in 200 towns around Europe, producing over 15,000 works. The main result was cheaper books. Many more people could buy them, and learn to read and write. Then they bought more books, read them, and learned even more. Schools and education changed forever.

Well I never!

• A printing system at the Lawrence Radiation Laboratory, California, can print the entire Bible in 65 seconds. That's 773,700 words!

GLASS-FIBER
OPTICAL CABLES

Glass-fiber cables are at the heart of today's communication revolution.
Thousands of miles of glass-fiber cables are being installed around the world, creating "data highways" along which a variety of information and services can travel. Glass fiber was first introduced to replace copper telephone cables. Capable of carrying 40,000 calls at the same time, the first long-distance glass-fiber telephone cable went into operation in 1983. Today, glass fiber carries far more than telephone calls. It is used to link up computer systems so that large amounts of data can be sent around the world, and to carry signals for cable T.V. Recent developments in technology mean that glass fiber can potentially carry 500 T.V. channels.

The cables (right) are made of fine fibers of glass. Although glass is usually a brittle, fragile material, the quartz glass used for the fibers is tough, flexible, and cheaper than copper. Many fibers are spun together to form cables. These are clad in a tough plastic sheath to protect them from damage.

The first transatlantic glass-fiber cable was laid in 1988, linking Britain and France with the United States. Special ships (left) are used to lay cables. The cables are played out slowly over the stern of the ship and come to rest on the seabed.

Glass fibers are used to link together computers so that they can communicate with one another, locally or nationally. Japan plans a glass-fiber system that will link virtually every home and office by 2015.

Undersea cables have to be armored to protect them from accidental damage. In shallow water they are buried to avoid damage from ships' anchors.

Cables are brought ashore at each end and are linked into the telephone network. Communications satellites share the market with undersea cables, but have yet to replace them.

G L A S S   F I B E R
H O W   I T   W O R K S

A caller's voice, picked up by a microphone inside the telephone, is converted into digital signals. These turn a laser on and off, sending light pulses along the glass fiber. Each fiber consists of a core of glass surrounded by an outer layer of glass with a lower refractive index. This reflects light at a slightly different angle and stops the light escaping from the fiber. The receiving phone decodes the digital signal. Computer information is transmitted in the same way.

Outer layer of glass

Light path

Glass core

The Starry SKIES

Like human beings, stars are born, grow old, and die. If we look hard enough, we can find stars of every age in the sky. Stars are formed from clouds of hydrogen gas collapsing under the force of gravity and turning into helium gas, in a process that produces huge amounts of energy. Near the end of the lives of giant stars, the helium changes into even heavier substances. Eventually these giant stars blow up in huge explosions called supernovae, scattering elements like carbon, silicon, iron, and oxygen into space. New stars and planets form from this debris. The Earth and everything in it, including ourselves, is in fact made of recycled material from a long-dead star.

GIANTS AND DWARFS
The biggest stars, or "red giants," have a lot of pressure in their core, burn quickly and brightly and die earliest, leaving the core as a "white dwarf." Tiny, dim stars, or "brown dwarfs," never become true stars. They get gradually fainter, and finally fade into "black dwarfs."

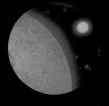

Into the void
After a huge star explodes, the core is left behind and collapses into a tiny point – a black hole. The pull of gravity from a black hole is so strong that not even light can escape from it.

THE LIFE OF A STAR

A star like the Sun begins as a cloud of gas and dust, which is gradually squashed by the force of gravity to make the star. At the end of its life it swells up into a "red giant" star, then puffs off its outer layers of gas into space. Even our Sun will finally end its life as a tiny "white dwarf" star.

A SCIENTIFIC BREAKTHROUGH
Sir Arthur Eddington (1882–1944) was the first person to realize that the mysterious spiral shapes seen in the sky were galaxies. He also proved that Einstein's theory of gravity was correct, by watching light being bent during an eclipse in 1919. Eddington wrote several famous books that explained the nature of the universe in a simple, understandable way.

OUR OWN GALAXY

The Milky Way is a spiral-shaped galaxy. Our solar system is on one of the "arms" of the galaxy, about two-thirds of the way out.

STAR PULSES

Super-dense stars called neutron stars, which measure about 20 miles across, spin quickly and send out radio signals. The regular pulses picked up from these stars by large radio receivers on Earth give them the name pulsars.

PATTERNS IN SPACE

Galaxies form in four different shapes. Spiral galaxies are like pinwheels, and the oldest stars are contained in elliptical (oval-shaped) galaxies. Barred spiral galaxies have a thick line running through the middle. Other galaxies have irregular shapes, depending on the number of stars and their position in space.

What would happen if I fell feet first into a black hole?

You would be stretched out like a piece of spaghetti, because the force at your feet would be stronger than that at your head. Then you would disappear beyond the "event horizon." Nothing, not even light, can escape from the black hole once it has passed this point.

THE CONSTELLATIONS

There are 88 identified constellations in the universe. Each has its own area of the sky, decided in 1930. The constellations are useful for finding your way around the sky.

THE BRIGHTEST STAR

Eta Carinae is the most luminous known star of all. It is 150 times bigger than the Sun, and six million times brighter.

THE FIRST MOTORISTS

After the experiments of the 1880s, the first car-making factories were set up in Germany and France during the 1890s. Engineers improved Daimler's engines, making them more powerful and reliable. Frenchman Emile Levassor was probably first to think of the car as a machine in its own right, and not just a cart without a horse. In 1891 he moved the engine from the back to the front, away from the mud and stones thrown up by the wheels. He also replaced the belt drive between engine and road wheels, with a clutch and a gearbox. The car as we know it today was quickly taking shape.

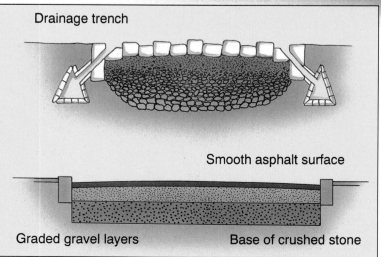

Napier 1913

Fun for the family
A trip in the car was an enjoyable outing, provided the rain stayed away. Early cars had no heaters and little protection from mud, dust, or the weather.

Roads to run on
Cars were faster than horse-drawn carts, so they needed better roads. Instead of packed-down layers of earth and stone (top right), engineers devised smooth surfaces of tarmac or asphalt.

Drainage trench

Smooth asphalt surface

Graded gravel layers

Base of crushed stone

A sign of status
Big houses, fine furniture and beautiful horses had been signs of wealth for centuries. Around 1900 a new symbol of status appeared: the car. As yet, the car was not a useful means of transportation. Roads were muddy, rutted cart tracks, and refueling places were scarce.

Flying the red flag

Under a law passed in Britain in 1865, steam traction engines were allowed to lumber along roads, provided they did not exceed 4 mph (6.5 kph) and a man with a red flag walked about 170 feet in front. The red flag was abandoned in 1878, but a footman still had to walk 60 feet in front. This law applied to any similar vehicle, including the first cars. In 1896 the footman was abandoned too, and the speed limit raised to almost 12 mph (20 kph).

Gas pump 1905

Buying a car

At first, only the rich could afford a car. But many people gathered in car showrooms to gaze at these newfangled pieces of machinery, which seemed to have little practical use.

Stopping for fuel

In the early days many car owners carried cans of spare fuel with them. There were no detailed maps and finding a gasoline station on a long journey was mostly a matter of luck.

The maker's name

Car manufacturers were soon competing to produce the best, fastest, or cheapest vehicles. Companies such as Buick and Austin designed easily recognized nameplates.

Advertising

The growing car business involved designers, engineers, manufacturers, mechanics, and of course advertisers. As roads became busier, they became valuable places for posters, advertising the latest cars to drivers.

HOW SOUNDS ARE HEARD

When sound waves enter our ears, they strike the eardrum which vibrates back and forth. This in turn causes tiny bones called ossicles to vibrate. These vibrations are turned by our "inner ear" into electrical signals that pass along nerves to the brain. When the signals reach the brain, we hear sounds. People cannot hear some sounds because they are too high or too low – not everyone can hear the high-pitched squeak of a bat. Dogs can hear higher pitched sounds than people, and a scientist called Sir Francis Galton (1822–1911) invented a whistle for calling dogs which was too high for people to hear. This kind of sound is called ultrasonic sound, or ultrasound.

WHY IT WORKS

The sound waves from your friend's voice make the plastic wrap vibrate. These vibrations are transmitted into your cardboard "ossicles" and can be seen by watching the mirror for movements. This is a simple model of how a real ear works.

Vibrations in the air (sound waves) enter the outer ear and make the eardrum itself vibrate. The three bones, the malleus, the incus, and the stapes, together known as the ossicles, transmit the vibrations through the middle ear

to the oval window, or vestibular fenestra. The force of the vibrations on the oval window is over 20 times greater than that of the original vibrations on the eardrum. The oscillations (or vibrations) of the stapes makes the fluid in the part of the inner ear called the cochlea vibrate. The cochlea also contains fibers that pick up the vibrations and send messages along nerves to the brain. Other parts of the ear control our balance – these are called semicircular canals.

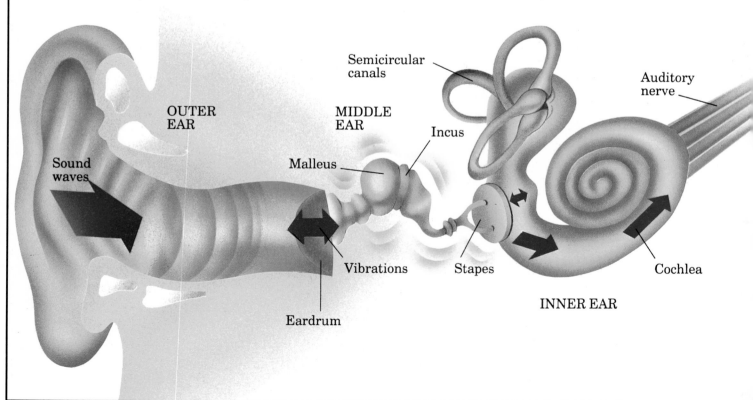

Semicircular canals

Auditory nerve

OUTER EAR

MIDDLE EAR

Incus

Malleus

Sound waves

Vibrations

Stapes

Cochlea

Eardrum

INNER EAR

HEAR THIS?

5. Shine a light onto the mirror and ask a friend to talk into the ear. Watch the mirror for vibrations.

5

1

1. To make an eardrum, stretch a piece of plastic wrap or a piece of an old balloon across the end of a tube. Fix in place with a rubber band.

3

2. Make a set of "ossicles" with two disks and a fork shape of thin cardboard held together with double sided tape.

2

3. Attach a small mirror or a disk of shiny foil to one end, and attach the other to the plastic wrap "eardrum" on the tube. Your middle ear is complete.

4. Make an outer ear from a cone of cardboard with a hole at its end. With careful use of pink tissue paper you can achieve quite a realistic look!

4

BRIGHT IDEAS

⚡ Watch the oval window of your ear again. How does it respond to shouting, whispering, whistling etc? The vibrations of your oval window could be a result of blowing on the eardrum, rather than the vibrations of sound waves. Use a radio to create sound without blowing.

⚡ A hundred years ago, people who suffered from hearing loss used ear trumpets. Find out about these devices! Can you make your own?

⚡ Listen to sounds blindfolded. Put your hand over one ear and try to tell which direction a sound is coming from. Listen to the same sound with both ears. Can you hear a difference? It is easier to hear the direction of sound with both ears.

WHAT IF AIRCRAFT DIDN'T HAVE WINGS?

Most of them would speed along the runway... and crash at the end, without taking off. Wings make a plane rise into the air, by providing lift. The wing shape also tells you how fast a plane goes. Slower planes, especially gliders, have very long, thin wings that stick out sideways. These wings have a very narrow chord (the distance from the front of the wing – leading edge – to the rear of the wing – trailing edge). This is the same design as the wings of gliding birds, like the albatross. Faster planes have swept-back wings, usually with a bigger chord, like fast hunting birds such as hawks.

An uplifting experience

Some aircraft don't get their lift from wings. It comes from a jet engine or propeller facing straight down, which pushes the aircraft into the air. A strange test craft from the 1950s, called the "Flying Bedstead," did this. So does the Harrier Jump Jet, which can take off straight up, and hover in mid-air. Hovercraft use lifting fans, like propellers facing downward, to create a cushion of air.

What if a plane flew upside down?

First, how does a wing work? Seen end-on, a wing has a special shape, known as the airfoil section. It is more curved on top than below. As the plane flies, air going over the wing has farther to travel than air beneath. This means the air over the top moves faster, which produces low pressure, so the wing is pushed upward – a force known as *lift* – which raises the plane. If a plane flew upside down, the wing would not give any lift. To overcome this, the plane tilts its nose up at a steep angle. Air hitting the wing then pushes it up, keeping the plane in the air.

Air moves faster

Lift

Wing

Air moves slower

Can planes fly in space?

A few can, like the X-15 rocket planes of the 1960s and the space shuttles of today. First, they need engine power to blast up there. The X-15 was carried up on a converted B-52 bomber, while today the shuttle uses massive rocket boosters. Once in space, there's no air (or anything else), so wings can't work by providing lift. The power for all maneuvers in space comes from small rocket thrusters.

How do spy planes fly so high?

Spy planes, such as the U-2 and the Blackbird, need to fly high, at 100,000 ft (30,000 m) or higher, so they are beyond detection by enemy planes or radar. However, as air is very thin at such heights, they need very special wing designs, with an extra-curved top surface, to give the greatest possible lift. With the arrival of spy satellites, the use of spy planes declined, until recently. A new generation of pilotless, remote-control spy planes, such as DarkStar, has arrived.

A QUESTION OF TASTE

BOTH NOSE AND TONGUE detect dissolved chemicals. In the mouth, these usually arrive in the form of flavor molecules from foods and drinks. In fact, it is believed that the multitude of flavors you experience when you eat and drink are based on various combinations of only four basic flavors – sweet, sour, salty, and bitter (see opposite).

Originally our sense of taste probably evolved to warn us about foods that were bad, rotten, or poisonous in some way. Then people learned what was good to eat, and how to recognize, cook, and flavor foods with a huge variety of sauces, herbs, spices, and other substances. Along with progress in agriculture and food storage, this turned eating from a necessity for survival into an enjoyable taste experience.

PAPILLAE
The surface of the tongue is covered with thousands of tiny pimples, called papillae (below left). They give the tongue a bumpy surface and make it rough to help move food around while you chew.

TONGUE NERVES
The tongue is connected to the brain by three cranial nerves (main picture). Two of these deal with the taste sensations, taking the information back to the brain where it can be processed. The third nerve controls the movements of the tongue, helping you talk and chew food.

POISONOUS PLANTS
Your sense of taste can protect you from substances that may be harmful to you if they got inside your body. Many poisonous plants, such as the Yew tree (left), have a bitter taste that will warn you of their dangerous properties.

PAPILLA

TASTE BUDS
Taste buds are tiny ball-like clusters of crescent-shaped cells, like segments in a microscopic orange (below). They are set into the surface of the tongue, especially around the sides of the papillae. A tiny opening lets dissolved flavor particles seep onto the hairs at the top of the taste bud.

Taste hairs

Taste bud

Taste cells

Nerve

Astronauts traveling in space have described a loss of taste as they circle the Earth (left). Because of the lack of gravity, excess blood flows to the head. This excess blood creates congestion, similar to a cold, which, in turn, diminishes the sense of taste. As a result many foods for astronauts are made extra spicy!

Cranial nerves

Bitter

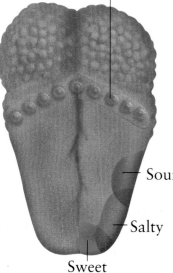

Sour

Salty

Sweet

TASTE AREAS
Most of the 10,000 taste buds are on the tongue, though some are on the rear roof of the mouth and in the upper throat. It is thought that different parts of the tongue detect different flavors (left). The tip picks out sweet ones, the front sides salty ones, the rear sides sour ones, and the rear center bitter flavors. The main upper surface of the tongue has few taste buds.

SPORTING TRUCKS

Custom trucks originated in the United States among drivers who owned their trucks. They tried to make their trucks look different from all the others on the road by painting them with startling designs and pictures. Customization may also include replacing some of the standard parts of a truck, such as the exhaust "stack" (a vertical exhaust pipe), and the fuel tank, with highly polished chromium plated parts. The demand for customized trucks is so great in the United States that many manufacturers now supply their trucks in a range of different color schemes. These serve as a starting point for the owner's unique "paint job."

Articulated truck tractors are mostly used to pull especially heavy loads along public highways at normal speeds, but the powerful tractors without their trailers are capable of traveling at very high speeds. Truck racing is one of the fastest growing motor sports in Europe.

Trucks have also taken part in ordinary car rallies including the annual Paris-Dakar race.

A truck "Superprix" race at the Brands Hatch circuit in England.

Truck racing using truck tractor units started in the United States, and then rapidly spread to Europe. There is now a European Truck Racing Championship. Every year, professional racing teams supported by many of the truck manufacturers compete for the title. The races test the trucks' top speeds and road-holding to the limit. The majority of the drivers still earn their living by driving ordinary trucks on the roads when they are not racing. Truck racing drivers do not yet receive the enormous amounts of money that Formula 1 racing drivers enjoy.

As with motor car racing, many of the improvements in ordinary truck design, engine efficiency and safety are made as a result of experimentation on the race-track.

A "funny car" with outsize wheels.

A Leyland Land train doing a "wheelie" at an exhibition event.

THE FINAL FRONTIER

"Have you come from outer space?" asked the farmworkers. "Yes!" came the reply.

When Yuri Gagarin landed in a remote field in Asia on April 12, 1961, he had fulfilled a dream that was as old as humankind. Although the official welcoming party wasn't there to greet him, the Russian cosmonaut Gagarin had been the first person to journey successfully into space.

SPACE RACE

Orbiting the Earth in *Vostok 1*, Gagarin's mission focused the Earth's attention on the possibility of exploring other worlds. Before the decade was over, NASA had put astronauts on the Moon. On July 20, 1969, a billion people watched in awe as men from the *Apollo 11* mission took their first tentative steps outside the Earth. Science fiction had at last become science fact.

Below *A two-seater* lunar rover *was folded and packed into one small cabinet on its trip to the Moon in the later* Apollo *missions.*

Left *Featuring aliens that live under the Moon's crust, Jules Verne's fantasy* From the Earth to the Moon *was written over 100 years before the real Moon landings.*

NO CHEAP THRILL

According to some estimates, an *Apollo* mission to the Moon today would cost a staggering $500 billion to fund. Scientists need to find cheaper ways to explore space. They have already sent robot probes to our solar system and beyond. These have brought back spellbinding images of our neighboring worlds. The farthest reaching probe, *Pioneer 10*, is now over 6 billion miles (10 billion kilometers) from Earth. That's roughly 100 times the distance from the Earth to the Sun. In October 1997, NASA launched a new probe, *Cassini*, that will take a closer look at the rings and moons of Saturn.

Right *Scientists on the space station* Mir *have studied how to support life for several months in space. This has helped them plan future missions.*

BACK TO THE MOON

Under its dusty surface, the Moon may be rich in valuable minerals. The Apollo astronauts only brought back 840lb (382kg) of Moon rock, but some scientists think that we should go back and mine the Moon.

NEW BICYCLES
CHANGING CONCEPTS

The shape of the bicycle, unaltered for nearly a century, has been transformed in the past ten years.

New materials, new designs, and wind tunnel testing have made bikes lighter, stronger and faster. Designing bikes is like designing aircraft: both have to minimize wind drag, maximize efficiency, respond quickly to the controls, yet be as light as possible. The key to the bike is the frame. Nowadays, the tubes in a racing-bike are made of aircraft-grade aluminum alloy, the gears of titanium, and flat disks are used instead of spoked wheels to reduce wind resistance. Brakes and gear mechanisms may be combined so that less time is needed to switch from applying the brakes to changing the gears.

The Lotus bike replaces the normal frame with a solid "wing" made of reinforced carbon-fiber. Designed for racing on a circular track, it has no gears or brakes. The rear wheel is a flat disk to minimize drag, but the front one is spoked. The flat handlebars allow the rider to lie almost horizontal.

Rider's seat

Rear wheel

British inventor, Clive Sinclair's Zike is a recent attempt to produce a light, powered bike. Batteries in the frame produce electricity to drive the Zike, which can also be pedaled. Designed for town use, the Zike has small wheels and a suspension system.

Rider's seat

Drive chain

Suspension forks

Handlebars are a key factor, because they control the position of the rider. The lower the rider, the less wind resistance; but careful wind tunnel testing is needed to create optimum airflow. Riders in races like the Tour de France use drop handlebars with a curved horseshoe-shaped bar on top, which they can tuck their elbows behind for sprinting. These became popular after American cyclist, Greg Le Mond, won the 1989 tour using them. Clipless pedals, which operate like ski bindings, are safer than the traditional toe clip.

The most expensive mountain bikes, like the Cannondale Super V, use air-sprung shock-absorbers to soak up the bumps. The movement of the front suspension can be adjusted.

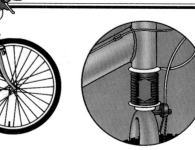

Solid "wing"

Three-pronged front wheel

MAVIC 3G

SAILING ON A BICYCLE
HOW THE LOTUS BIKE WORKS

Most bikes lose speed in a crosswind. That is because drag increases sharply when the air is flowing past them at an angle. The Lotus bike, by contrast, is designed to go faster in those conditions by using its flat frame as a sail, taking advantage of the wind.

To achieve this result, the bike was put into a wind tunnel with its rider, Chris Boardman, in the saddle, and wind resistance was measured at different angles. By adjusting the shape and curvature of the frame, and ensuring that it and the solid rear wheel acted as a unit, it was predicted that on a circular track, a fraction of a second would be gained every lap. Boardman went on to win an Olympic Gold at Barcelona. The precise position of the rider, allowing air to flow between him and the bike, was also perfected.

The world's oddest bike is Behemoth, designed by Steve Roberts. It has 105 gears, carries four computers, a satellite navigation system, a refrigerator and solar cells to power them all. Behemoth – it stands for Big Electronic Human-Energized Machine Only Too Heavy – is a mobile office on which Roberts has pedaled 19,311 miles across the United States.

BLACK HOLES · Doors to other universes or deadly dead ends?

Above *Cosmic researcher, Stephen Hawking, thinks the universe is peppered with black holes of all sizes, large and small.*

When we launch a rocket into space, we need to make it move fast enough to escape the pull of the Earth's gravity. If Earth were heavier or denser (more tightly packed), our rocket would need to go faster. Now imagine an object that is so heavy and dense that not even light — the fastest thing in the universe — can travel quickly enough to escape it. An object like this is called a black hole.

DEAD STARS

Black holes aren't just science fiction — scientists think they really do exist. They could be formed when massive stars die, for example. Dying stars collapse in on themselves to leave super-heavy, super-dense remains. The remains could become so tightly packed, they form a black hole.

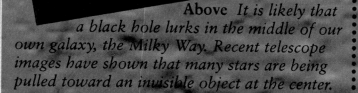

Above *It is likely that a black hole lurks in the middle of our own galaxy, the Milky Way. Recent telescope images have shown that many stars are being pulled toward an invisible object at the center.*

TOMORROW IS YESTERDAY

Some theorists have imagined how black holes could work like time machines. The gravity around a black hole could be so great that it would cause time and space to fold back on itself, creating a loop. If you traveled into this loop, you could find that time doesn't go forward as usual, but doubles back on itself. So you'd end up going back in time!

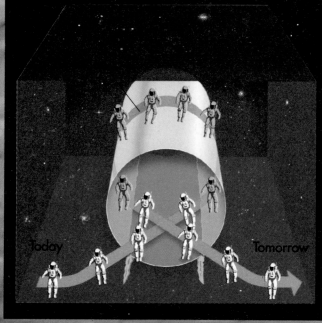

IN THE DARK

We will never be able to see a black hole as light cannot escape from it. But we can spot a black hole in action. As it creates a huge gravitational force, it pulls gas clouds and other stars toward it. Matter spiraling into a black hole heats up, giving off X rays that we can detect.

People have come up with lots of imaginative ideas about travel through black holes. Some suggest they could be gateways to other times or universes (see box *above*). But Stephen Hawking offers a word of caution. He doesn't think anyone could survive such journeys, even if they were possible.

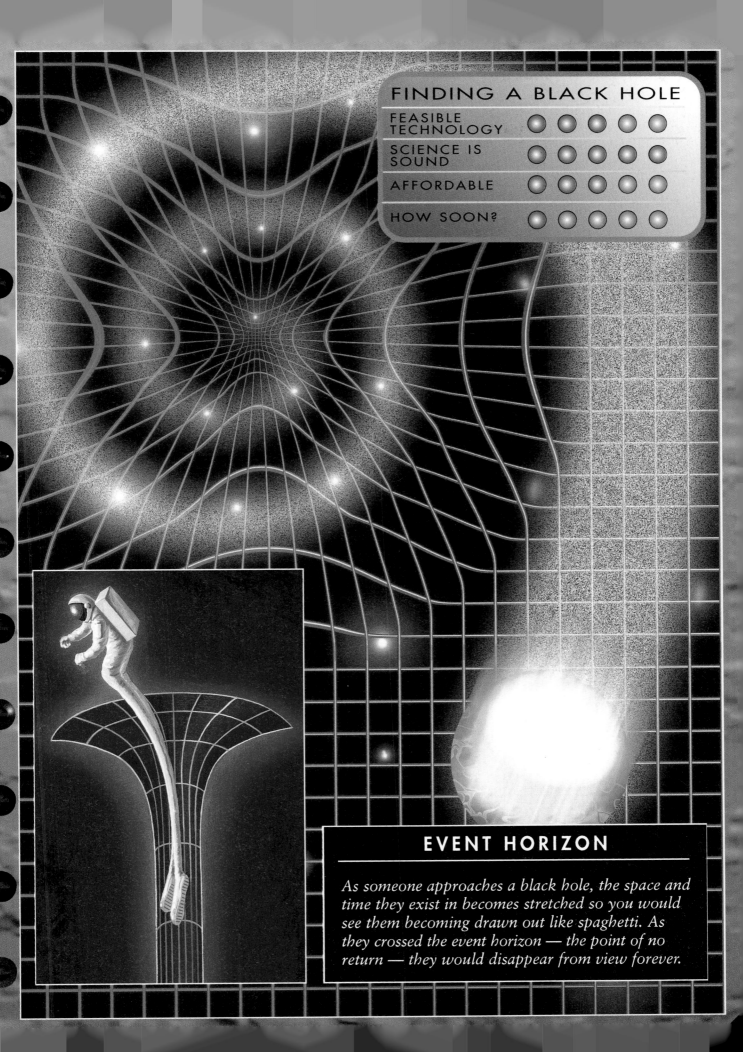

FINDING A BLACK HOLE

FEASIBLE TECHNOLOGY	○	○	○	○	○
SCIENCE IS SOUND	○	○	○	○	○
AFFORDABLE	○	○	○	○	○
HOW SOON?	○	○	○	○	○

EVENT HORIZON

As someone approaches a black hole, the space and time they exist in becomes stretched so you would see them becoming drawn out like spaghetti. As they crossed the event horizon — the point of no return — they would disappear from view forever.

4 SENDING SOUNDS TO THE BRAIN
*As the membranes bend with the pressure
ripples, the hairs vibrate. Their
movements turn into nerve signals
that pass along the cochlear
nerve to the brain
(below).*

OUTER
HAIR
CELLS

INNER
HAIR
CELLS

COCHLEAR
NERVE

3 UP THE COCHLEA
*The cochlea is a coiled
tube that spirals for two and
three quarter turns. Inside, the tube is divided by a Y-shaped
membrane (above). Ranged in rows along the bottom of one
arm of the membrane are about 24,000 hair cells, arranged in
two groups – outer and inner hair cells. Each hair cell has up
to 100 hairs. Incoming sound energy travels up the cochlea,
vibrating the membranes as it goes. Once it reaches the peak,
it travels back down the cochlea where its energy disperses.*

INSIDE THE EAR

Like all other sense organs, the ear converts one type of energy – in this case, the energy of sound waves – into tiny electrical nerve signals. This transformation happens deep inside the ear, in the snail-shaped organ of hearing called the cochlea. This grape-sized part is embedded and protected inside the skull, just behind and below your eye.

The whole of the ear is divided into three main parts – the inner, middle, and outer ear. The outer ear is made up of the funnel-shaped piece of flesh that sits on the side of your head and the ear canal. The middle ear consists of the eardrum and the three ear bones. Finally, the inner ear is made up of the cochlea and the cochlear nerve.

Should the cochlea or ear bones become damaged, hearing may be lost. Doctors can restore hearing by using an artificial implant (above). This has a microphone that converts sounds into electronic pulses. These are sent down a wire into the cochlea, where they stimulate the cochlear nerve.

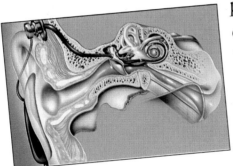

2 INTO THE COCHLEA
The base or foot of the stapes is attached to a thin, flexible part of the cochlear wall. As the stapes tilts and pushes in and out like a piston, it transfers waves or ripples of pressure along the fluid inside the cochlea.

STAPES INCUS MALLEUS EARDRUM

COCHLEA

1 PICKING UP SOUND
The eardrum is a thin, tight sheet of skin-like material. Sound waves bounce off it, making it vibrate. These vibrations are passed along the three ear bones, or ossicles: malleus, incus, and stapes. The bones pass the vibrations to the cochlea.

FIGHTING SHIPS

Fighting ships were conceived as fighting platforms for troops to board and capture the enemy vessels. The early Egyptian galleys were probably the first real fighting ships, and although sails were added through the ages, galleys with banks of oarsmen were still in service until the 17th century. The great ships or galleons like the *Mary Rose* (which sank because she was made top heavy) were followed by the ships-of-the-line, like *Victory*. They used their broad sides of cannon fire to cripple opposing ships before boarding and capturing the enemy. The French *Gloire* (1859) was the first warship to carry iron armor plates, followed in 1860 by the English all-iron, screw-driven *Warrior*.

Battleships

In *Warrior* (below), iron replaced wood, steam replaced sail, and guns replaced cannon. *Dreadnought* was one of the first new class of

battleships. It was launched in 1906, had ten 12-inch guns and a crew of 800 men. It was the first steam turbine-powered battleship.

HMS *Dreadnought*

Triremes

The trireme (above) appeared in the Mediterranean around 500 B.C. A typical Greek trireme was about 130ft long and 20ft wide. It had three banks of oars operated by 170 oarsmen, and also carried warriors for combat. It attacked and sank other ships by ramming.

Fire control radar

Bridge

SAM launcher

Automatic gun

Aircraft carriers

The first aircraft carriers were introduced between the two World Wars and during World War II they were the most important ship type.

Modern navies are based around aircraft carriers. The biggest carriers in the world are the *Nimitz* (below) class of the U.S. Navy. Their flight decks are 1080 feet long and they carry about 90 aircraft and helicopters.

New Developments
High-tech developments are likely in naval craft. The U.S. Navy's *Sea Shadow*, shown left, can avoid detection by radar.

Gunboat diplomacy
One of the main jobs of today's fighting ships is to wait near political trouble spots. This military "show of strength" known as "gunboat diplomacy" can be used to support or threaten. U.S.S. *Nimitz* is seen here displaying its airpower.

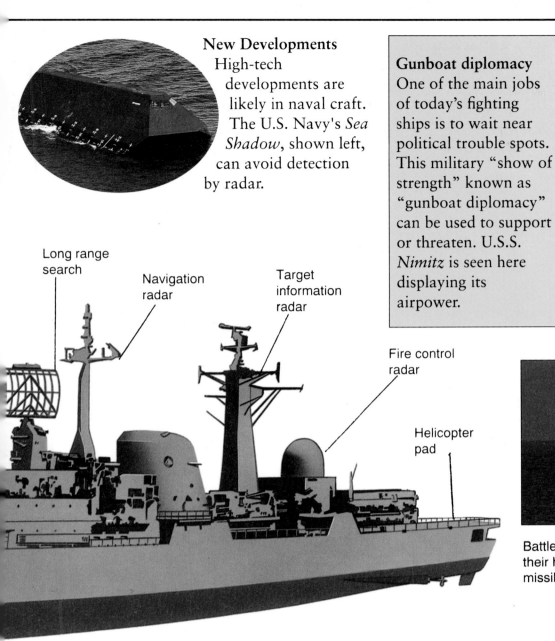

Long range search

Navigation radar

Target information radar

Fire control radar

Helicopter pad

HMS *Glasgow*, a Sheffield (type 42) class destroyer.

Battleships in the Gulf conflict used their huge guns and launched cruise missiles.

The modern ship
The use of iron and explosive shells revolutionized battleships. Modern fighting ships, such as the destroyer above, are fast, light, and crammed full of electronics. The gas turbine engines give it a top speed of 29 knots. Armament includes light guns, torpedoes, and surface-to-air (SAM) missiles for attacking aircraft. These are controlled by the computerized fire-control system. A helicopter is carried for antisubmarine operations. In a conflict, the ship's main role is to defend the fleet's aircraft carriers.

High-speed boats
Some countries operate high-speed patrol boats around their coasts. They often include hydrofoils and hovercraft as well as gas turbine boats. Iranian gunboats sank several oil tankers in the Gulf with torpedoes, during the Iran/Iraq war of 1984/1985.

MAGNETIC ATTRACTION

Archaeologists and beachcombers often make use of metal detectors to locate buried objects or treasure. Materials are either magnetic or nonmagnetic. Most, but not all, metals are magnetic. Iron has the strongest magnetic attraction. Nickel and cobalt are also magnetic, as are the alloys, or mixtures, of these metals. Aluminum, copper, and gold are nonmagnetic. Magnetic ferrites (metals containing iron) can be used to make hard magnets, like the refrigerator magnets pictured here. These are known as permanent magnets. Soft magnets are temporary and are easy to magnetize and demagnetize. Magnetic materials can be easily separated from other materials. When aluminum cans are recycled, they are sorted from other metals by using a magnet. You can make your own metal detector with an ordinary magnet. See if you can find any buried treasure!

SECRETS IN THE SAND

1

2 Hold the cone in place by attaching small strips of paper to it and winding them tightly around the stick. Now decorate the cone with paint or colored paper.

1 Make a circle, 4 inches across out of colored cardboard and cut a slit from the edge to the center. Overlap the two ends and glue, to create a flat cone. Push one end of a stick through the center and attach a button magnet to the stick with clay.

2

4 Move your metal detector slowly above the surface of the sand. Try it at various heights. You will soon discover how low you must hold it to attract objects.

3 Half fill a shallow container with clean sand. Bury a variety of objects in the sand, metals and nonmetals.

3

WHY IT WORKS

A magnet exerts a force on a nearby piece of magnetic material by turning it into a weak magnet – this is magnetic induction. A magnet is made up of many tiny parts called domains. Each one is like a mini-magnet, and they all point in the same direction. The domains in a metal are jumbled up. When a magnet comes into contact with the metal, the domains line up and the metal becomes magnetized. A strong magnet can act over quite a distance. Each object picked up from the sand is a temporary magnet because the domains inside become aligned.

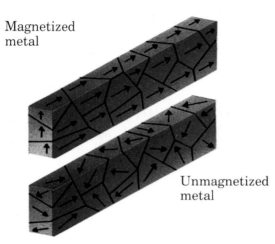

Magnetized metal

Unmagnetized metal

BRIGHT IDEAS

☀ Predict which objects you expect your magnet to pick up - you may be surprised! See how near to the sand the magnet must be held before it picks anything up. At what height does it fail to attract any of the hidden objects? Keep a record of your results.

☀ Which objects does your magnet pick up? Which are left buried in the sand? What does this tell you about them? (Hint: the answer is on this page.) Notice whether any of the magnetic objects keep their magnetism and attract other objects.

☀ Try a different kind of magnet. See if you can pick up any more objects with it. See what happens if you add more sand to the container.

☀ Find out which other metals are non-magnetic. Collect some empty drink cans and sort them with a magnet. Remember, aluminum is nonmagnetic. Save the cans for recycling.

4

THE REUSABLE SPACECRAFT

Conventional rockets are used just once, then thrown away. The space shuttle is different; it takes off vertically, like a rocket, enters space as a spacecraft, and then returns and lands on a runway like an aircraft. The space shuttle is scheduled to carry most of the parts of the International Space Station currently being assembled in space. To put a satellite into orbit with NASA's space shuttle costs up to $250 million, no less than a conventional rocket. The popular dream of ordinary people paying for a space ride is still many years away.

Development
The design of the shuttle drew on experience from a series of rocket planes developed in the United States. The first of these, the Bell X-1, launched in mid-air from beneath a B29 bomber, was the first aircraft to exceed the speed of sound in 1947. Later models (below) showed the rounded shape and V-shaped delta wings of the shuttle, designed to resist the intense heat of reentry and then to glide swiftly to a landing.

X-15
The Bell X-15 rocket plane (above), tested in the 1960s, reached speeds of more than 4,000 mph and attained heights of 67 miles, the very edge of space.

X 24A

M2F3

X 24B

The shuttle
The shuttle is built of aluminum alloy, covered with ceramic tiles to protect it from the heat of reentry. The cargo bay is 60 feet long by 15 feet wide, which is about the size of a railway freight wagon. The doors are made of carbon-fiber reinforced plastic. The stubby wings allow the shuttle to glide, though very fast, and land at more than 200 mph The flight deck is the upper level at the front, with the galley and sleeping berths below in the mid deck area. Each shuttle costs about $1.1 billion.

Satellite payload

Fuel tanks

Payload handling controls

Airlock

Oxidizer tank

First flight

The space shuttle Columbia lifted off for its maiden flight in 1981. In general, the shuttle program has been successful. It launches satellites regularly, carries out experiments in space, and also does secret military work. It has made dramatic rescues.

Flight plan

The flight sequence of the shuttle appears above. For lift-off (1), the shuttle uses its three main engines, plus two boosters. Extra fuel is carried in a huge internal tank. After two minutes, the boosters burn out and parachute into the sea (2). Six minutes later, the main engines stop, and the fuel tank is released (3). The final step into space is made by smaller orbital engines (4). After landing (5), a Boeing 747 returns the shuttle to the launch pad (6).

Space radiator

Forward control thrusters

Nose wheels

International rescue

In 1992 (below) three shuttle astronauts spent more than eight hours on a space walk, wrestling a four-ton communications satellite, Intelsat-VI, into the cargo bay. There they fitted a new rocket motor and sent the satellite off on its true orbit, 22,300 miles above the Earth.

Disaster

In America's worst space disaster, hot gases leaked through a joint in the booster casing during the launch of Challenger in 1986. A tongue of flame burned into the main tank and ignited the fuel, blowing the shuttle to pieces and killing its seven crew members, among them Christa McAuliffe, a teacher. The disaster set back the program by nearly three years, as engineers struggled to prevent it from ever happening again.

INDEX

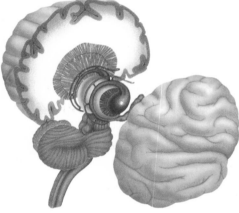

INDEX